THE HUMAN STORY

Where we come from and how we evolved

CHARLES LOCKWOOD

Foreword by Chris Stringer

Published by the Natural History Museum, London

First published by the Natural History Museum, Cromwell Road, London SW7 5BD

Reprinted with updates 2008

ISBN 13: 978 0 565 09214 6

A catalogue record for this book is available from the British Library.

Edited by Karin Fancett
Designed by Mercer Design
Reproduction by Saxon Photo Litho
Printing by C&C Offset

Front cover: Only about 2000 generations have passed since *Homo neanderthalensis*, Neanderthal man, became extinct. © Michael Long/NHMPL
Back cover: The incredible skeleton of *Australopithecus afarensis*, Lucy, uncovered after 3 million years. © Natural History Museum, London
Spine: Skull of *Homo heidelbergensis*. © Natural History Museum, London

FOREWORD

In 2008 we celebrate the presentation, by Charles Darwin and Alfred Wallace, of the ideas on evolution to the world. Then, the first fossil human finds were only beginning to be made or recognized, and evolutionary ideas, palaeontology and archaeology were still in their infancy. Now, there is a rich and ever-growing record from Africa, Asia and Europe, and a plethora of accompanying scientific names, some of which are difficult to articulate, even for the specialists! In this book, Charles Lockwood guides us expertly through the evidence for over 20 of the fossil species that represent our ancestors or close relatives.

New finds from various regions of Africa are fleshing out the possible beginnings of the human evolutionary line, with several candidate species vying for the title of our earliest ancestor. The surprisingly early spread of humans from Africa is well illustrated by the unexpectedly rich site of Dmanisi in Georgia, dating close to 2 million years ago. Here, skulls, jaws and parts of skeletons of some primitive human fossils have been found beneath the remains of a medieval village. The small brain size and very basic stone tool kit of these early humans are surprising; many experts assumed that significant advances in intelligence and technology would have been necessary, for these early humans to make the first moves out of their ancestral African homelands. Indeed, the evidence is so challenging, it has even been suggested that we rethink the idea that the genus *Homo* originated in Africa!

Fossils ranging from South Africa to Australia are revealing the origin and spread of our own species from Africa. How little we yet know about later human evolution, in regions such as Southeast Asia, has been highlighted by the discovery of a remarkable skeleton of a very primitive, human-like form on the island of Flores, Indonesia. The existence of this creature, given the name *Homo floresiensis* and nick-named the 'Hobbit', was completely unsuspected. The fact that it apparently survived until less than 15,000 years ago means that modern humans, dispersing through the region to Australia, must surely have encountered these strange relatives. The evidence is so challenging to conventional thinking that some experts have questioned its reality, insisting that the unusual features of the material are the products of abnormality rather than evolution.

As Charles explains, despite all the new finds and advances in study techniques, there are still many fascinating puzzles about our evolutionary origins. The exact nature of the last common ancestor we shared with chimpanzees remains uncertain, as is the date at which it lived, and the environment in which it originated. There are also many different ideas about why our ancestors began the fundamental human adaptation of walking upright on two legs, but we are still far from knowing whether any of them are correct. And although most experts believe the event happened in Africa prior to 2 million years ago, we do not really know when, where and why the first members of the genus *Homo* evolved. Equally, for *H. sapiens*, what processes were involved in the origin of our species, and how did we come to replace those other surviving species, such as the Neanderthals and the strange 'Hobbit' from Flores? Further finds will help us to close many of the gaps in our knowledge, but in the meantime, Charles's book provides a readable, up-to-date, well-illustrated and user-friendly summary of the evidence as it stands today.

Sadly, Charles died in a traffic accident in London on July 14th 2008. He was already an established figure in research on human evolution and was about to take up the position of Director of the Institute for Human Evolution in South Africa. He will be sorely missed.

Professor Chris Stringer
Research Leader in Human Origins and Director of
Ancient Britain in its European Context Project
Department of Palaeontology
Natural History Museum
London

INTRODUCTION

Here's a tongue twister for you: *Australopithecus anamensis* was an ancestor of *Australopithecus afarensis*. It is difficult to say even once. But this tongue twister means something – it refers to two of our earliest ancestors who existed long before written history, before humans spread throughout the world, and even before human-like creatures made their first tools from stone. The record of human evolution extends back more than 6 million years, to the point when our lineage split from the chimpanzee lineage.

We are lucky to have an excellent record of bones and stone tools that tell us where, when, and how humans evolved. The record is so extensive that scientists have now named more than 20 species in this 6-million-year span. Each of these species was a group of individuals who were genetically closely related and who bred with one another. Humans are a species – *Homo sapiens* – and so are chimpanzees (*Pan troglodytes*) and gorillas (*Gorilla gorilla*). When we talk about human evolution, it is important to name species so that scientists can communicate about the groups of fossils associated with that species, determine how they behaved, and find out how they are related to us.

All of the species that are more closely related to humans than to any of the living apes are referred to as 'hominins', members of the taxonomic group Hominini. In this book, each of the hominin species is described roughly in order of when it lived, grouped into five main sections: earliest hominins, *Australopithecus*, *Paranthropus* (or robust *Australopithecus*), early species in the genus *Homo*, and finally the later *Homo* species – including us, *Homo sapiens*. The complete human story emerges through the book, but it doesn't have to be read in order. If you want to know, for example, what *Homo habilis* was, you can turn to that section of the book and find out. Here we introduce the characters that play out the human story, a story that involves a large and diverse cast.

Evidence from bones

Skeletons of three hominin species: *Homo erectus*, *Australopithecus afarensis*, and *Homo sapiens* (from left to right).

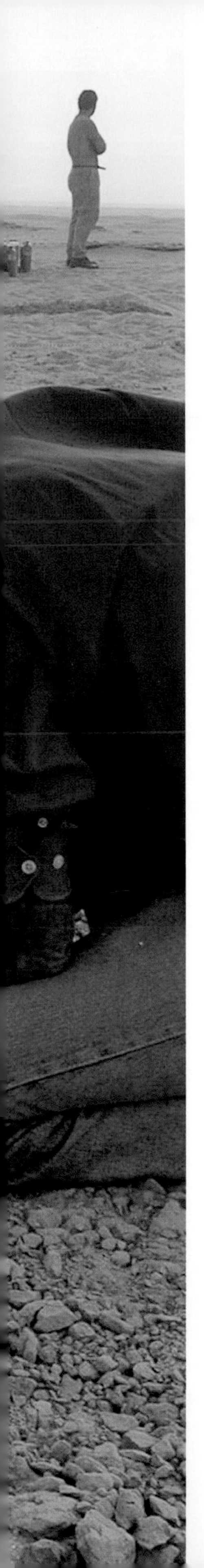

EARLIEST HOMININS

Genetic comparisons of living humans and apes tell us that our closest relative is the chimpanzee, and that we split from the chimpanzee lineage between 6 and 8 million years ago. For a long time in the study of human evolution we had no fossils that could tell us about the earliest hominins. Recently that has changed. Here, the species have all been described as ancestors to later hominins, but scientists are not entirely certain about the role each species plays in human evolution. The different names show the uncertainty – there is limited evidence with which to compare the species directly and determine how many distinct groups are represented. It will take time to know exactly how to interpret the fragmentary fossils, and the names may change too. Scientists may decide to lump the various discoveries into a smaller number of species. Despite this we can still ask, "What were the earliest hominins like?"

Searching for the most ancient hominins
Palaeontologists working in the Djurab desert of northern Chad, where *Sahelanthropus* was found. These sites are between 6 and 7 million years old.

Identifying the first hominins

Before we even look at the fossil record, we can make predictions about what the first hominins would look like. In the ancestor that we share with apes, we would expect to see anatomical features and behaviours shared by chimpanzees, gorillas, and orangutans – the living great apes. Starting from that ancestral point, the first hominins must have evolved at least one feature that we see in modern humans. They would not be expected to look just like us, but would show at least one characteristic that places them on the human lineage.

In looking for these features scientists focus on two major areas: anatomy related to walking upright on two legs – bipedalism – and the size and shape of the canine and first premolar teeth. The fossil record shows that some other important human characteristics, such as large brains, come much later in time, so these features are less important in identifying the first hominins. Humans are the only primates to routinely walk bipedally, and therefore bipedal walking is an obvious feature to look for in comparing hominins to other species. Many other primates walk bipedally for short periods or stand upright to obtain food, but no other species moves this way as its principal way of getting around.

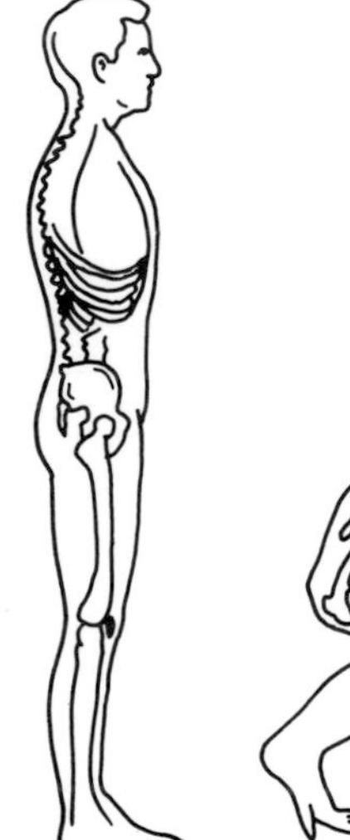

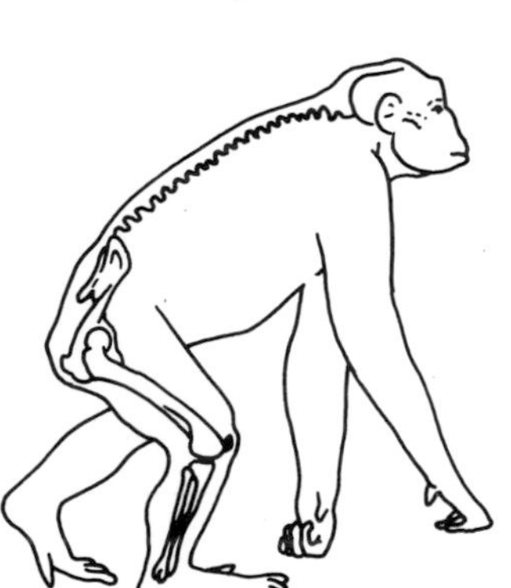

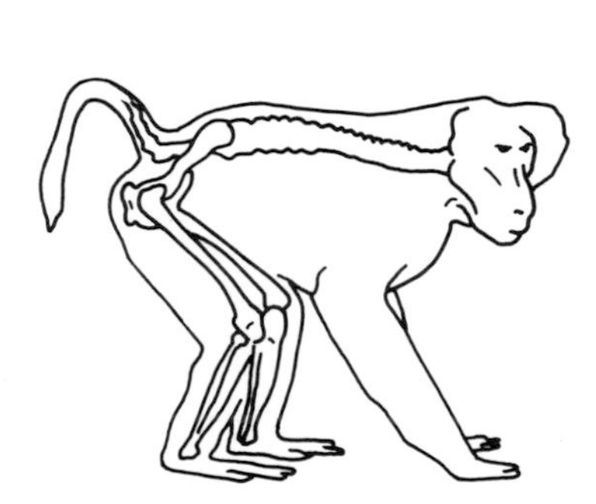

Standing up

Unlike apes and monkeys – shown here by a chimpanzee and a baboon – humans are adapted to walking upright. This impacts our skeleton in many ways. For example, curvature of the lower back allows us to keep our trunk over our hips without much effort.

Differences between humans and apes in their front teeth appear to be more subtle, but they are just as important as bipedalism in identifying the first hominins. Humans have smaller canine teeth than apes and monkeys. Because our canines are small, the lower first premolar has changed as well. In apes and monkeys, this

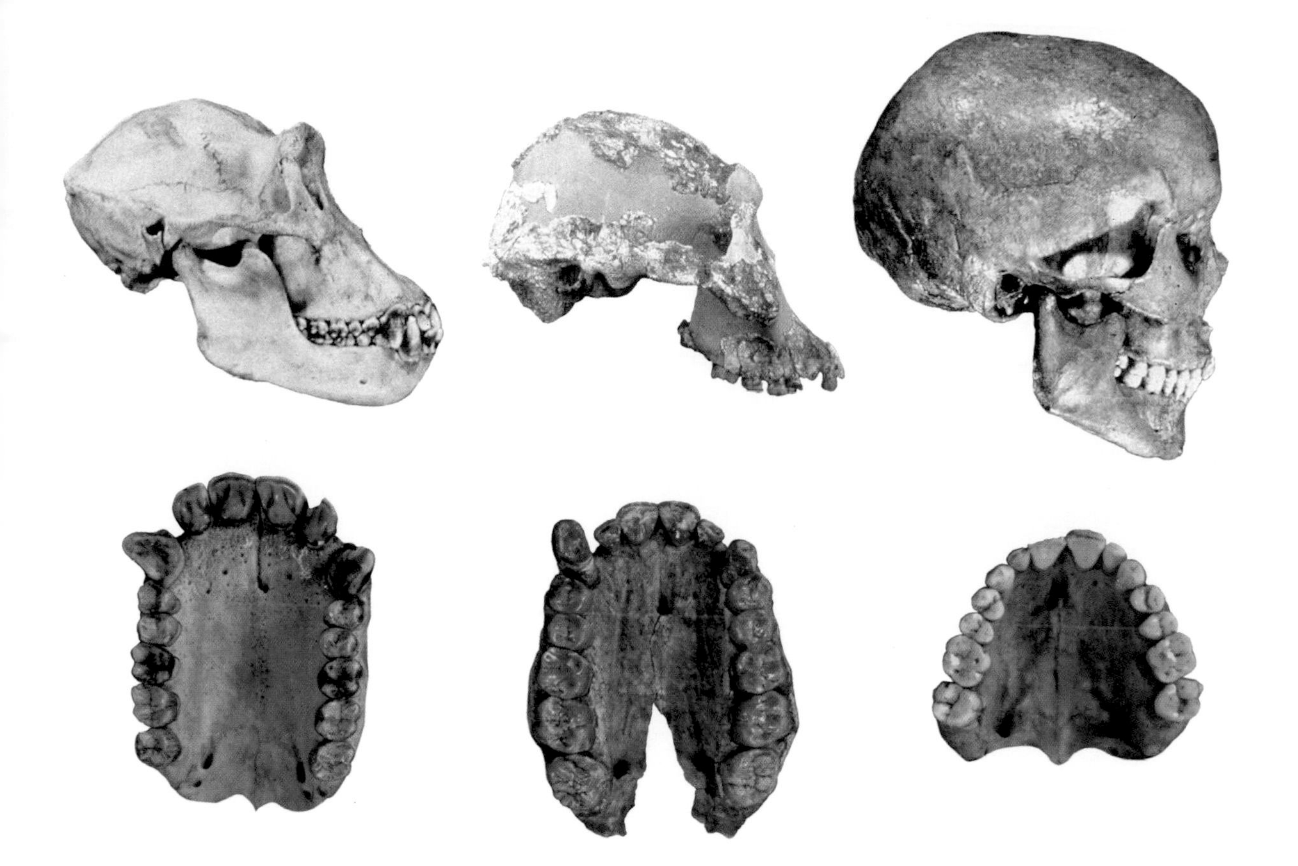

tooth has a single point (cusp) and a long 'honing' surface – it sharpens the fang-like upper canine as the jaws interlock and the two teeth wear against each other. In us, it has two cusps and is completely different in shape.

Going from a large, fang-like canine to a tiny one clearly requires intermediate stages, and for the most part the fossil record shows this transition well. The earliest hominins are called hominins because they show a number of small changes in the canines and premolars, and it appears that their lower first premolar is no longer used for honing. Of course, the canines of early hominins are still much larger than ours! It is only in comparison to the apes in general that their canines are called 'small'. Deciding whether fossils are hominin depends on the results of detailed comparisons of key features such as these.

Hominin anatomy

The skull and upper jaw of *Australopithecus afarensis* (see p.28) are shown here between a chimpanzee and a human. Chimpanzees have large canine teeth for use as weapons. Canines are smaller in fossil hominins and humans.

Sahelanthropus tchadensis

When the discovery of *Sahelanthropus* was announced in 2002, Bernard Wood remarked in *Nature* that it "fundamentally change[s] the way we reconstruct the tree of life". Palaeoanthropologists such as Wood were astounded at the location, the date, and the appearance of this new species. The site in Chad, in central Africa, is far to the west of most early hominin sites, and animals found with *Sahelanthropus* suggest it is around 6 to 7 million years old. This makes it the oldest species yet found that may be a hominin. When you first look at the face of the skull, you see a modern-looking form with a hulking browridge, not unlike *Homo erectus* – a much more recent species described later in the book. But the rest of the *Sahelanthropus* skull is decidedly more ape-like than other hominins. Hence, the conundrum for scientists: where does 'Sahel man' belong in the human family tree?

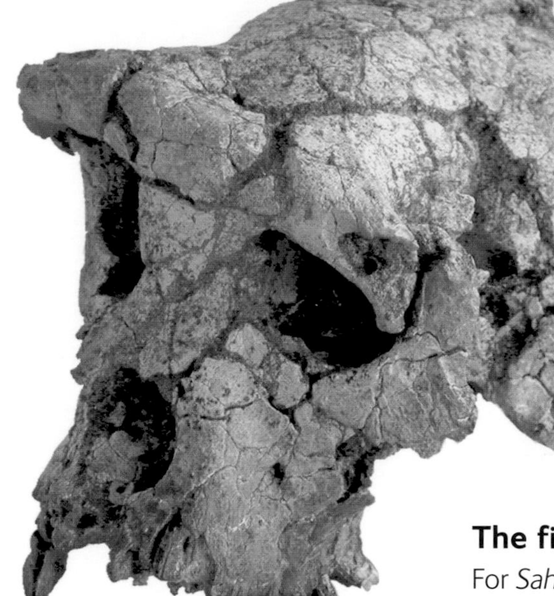

Oldest species

Nicknamed 'Toumai' – or 'hope of life' – this remarkable skull is between 6 and 7 million years old and is the most complete specimen of *Sahelanthropus*.

The first biped?

For *Sahelanthropus*, no bones of legs and feet have been found to determine if it was bipedal. However, the position and orientation of its foramen magnum – the hole in the base of the skull where the spinal cord passes through – suggests a posture similar to bipeds. It is horizontally oriented and relatively far forward, just what is needed if you are walking upright and looking straight ahead.

Sahelanthropus also may have had small canines and a hominin type of premolar – one that does not sharpen the upper canine. These changes in the front teeth are just as important as bipedalism in identifying the first hominins.

Lifestyle

It is likely that *Sahelanthropus* ate mainly fruit, with smaller amounts of a variety of other foods as well. But we need more evidence to reconstruct its diet in detail. If *Sahelanthropus* was bipedal, that would have affected the way it moved around the landscape and the way it obtained food, but apparently the types of food were not that different from the range of foods eaten by modern apes.

An important but unresolved question with *Sahelanthropus* is the exact nature of its habitat. The key specimens were found with a variety of habitats close to each other, ranging from desert conditions to the thick forests surrounding rivers. It is difficult to tie the species to one of these environments, although it is intriguing that from this early time period, open habitats such as grasslands are found nearby hominins, and these contrast with the forested habitats of modern apes.

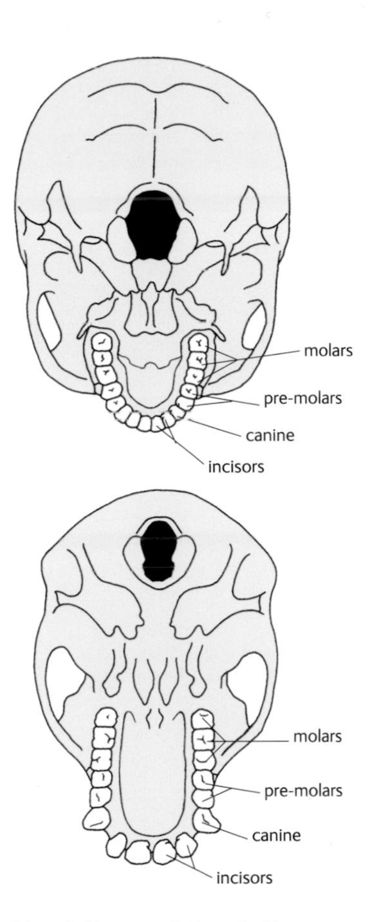

Bipedalism and the skull

The large hole in the base of the skull is where the spinal cord enters. In humans (top), it is further forward than in the apes (bottom), because of our upright posture.

Orrorin tugenensis

Sahelanthropus may be the oldest species discussed in this book, but the first discovery that broke the 5-million-year mark in human evolution is a species from Kenya. It was called *Orrorin tugenensis*, which means 'original man' in the local language of the Tugen Hills, the region where it was found. Again, its date of 5.8 million years is in the general time range that geneticists consider to be the time of origin of the human lineage. The date is reliable, as it comes from absolute dating of volcanic layers (lava flows) above and below the fossils.

Tugen Hills, Kenya

Teeth, jaws, and limb bones of *Orrorin* were discovered in this region in the year 2000.

Legging it

Orrorin is unusual as it is known mainly from parts of its skeleton other than its skull. There are fragments of jaws and teeth, but the main evidence about *Orrorin*'s lifestyle comes from two thighbones, especially the parts that connect with the hip. These thighbones (femurs) may be the oldest direct evidence for hominins walking on two legs.

Walking upright requires a substantially different hip joint to that seen in apes. In humans, the joint is stabilised by ligaments running across the joint between the pelvis and the femur. The marks left by these ligaments are clearly visible on most human femurs but not in most apes. When such marks were observed on *Orrorin* fossils, part of the case for bipedalism was made.

Another key difference between humans and apes is the structure of bone inside this part of the femur. With each step, humans bear all of their weight through this one joint, and nearly all the stress is placed on the bottom edge of the femoral neck – the narrow structure between the long part of the femur and its round 'head' (the ball part of the ball-and-socket hip joint). As a result, bone is denser in the bottom part of the human femoral neck, while in apes the hard bone is spread evenly around the edges.

In well-accepted bipeds such as *Australopithecus afarensis* (see pp. 22–30), the distribution of this hard bone is often cited as good evidence for bipedalism. It can be seen either in naturally broken

Signs of walking

Later bipeds, such as *Australopithecus afarensis* (middle), have a characteristic pattern of bone distribution in the neck of the femur (thighbone) that is similar to humans (right) and unlike chimps (left). Bone is more dense at the bottom edge of the femoral neck, because this area is subject to the most stress when walking upright.

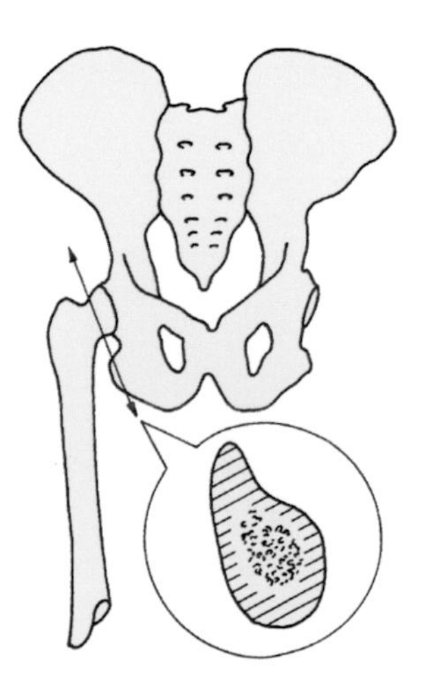

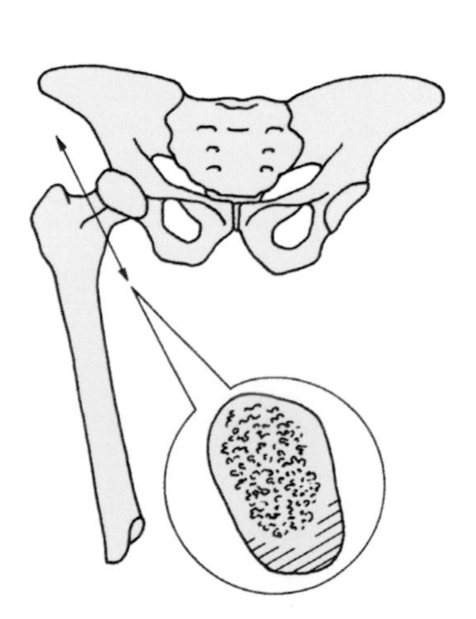

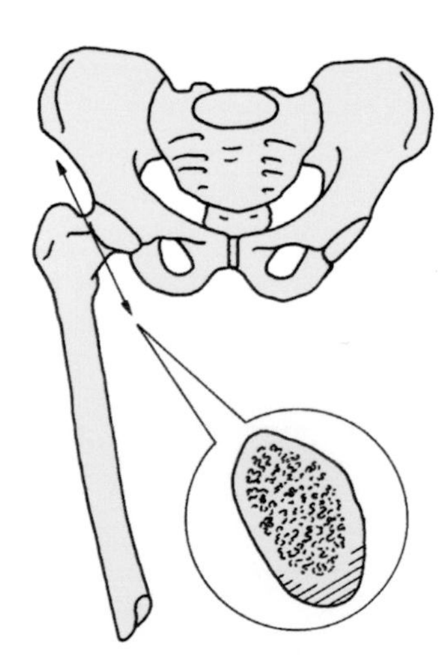

fossils or in specimens that have been scanned with medical imaging equipment. With *Orrorin*, the same pattern is more subtle, resulting in some disagreement about whether this is valid evidence for bipedalism.

A different picture from the teeth

Orrorin is different from *Sahelanthropus* in that it has ape-like canines and premolars. The primary evidence for it being a hominin, however, is in the leg bones. Its discoverers also pointed out that *Orrorin* has thick tooth enamel, like humans. However, since some other apes, such as orangutans, have thick enamel as well, interpreting this feature in *Orrorin* is not straightforward. As you can see, different kinds of evidence have been used to justify each of these species – *Sahelanthropus* and *Orrorin* – as a hominin, and it may be that only one of them (or neither) is actually part of our ancestry.

The dating game

To work out which is the first hominin, we need to know how old these discoveries are in geological terms. Dates for *Orrorin* and *Ardipithecus* fossils are reliable, because they come from 'absolute' dating techniques. The most common techniques are radio-isotopic methods, making use of the process of radioactive decay and the fact that this process behaves more or less like a clock.

We don't always know the dates so well, because these clocks have to be set by particular conditions. Eastern African hominin sites are a perfect setting, because volcanic activity was common through the period during which fossils were deposited. Volcanic eruptions 'reset' a process of decay for an isotope of potassium to argon, because the high level of heat eliminates argon gas, and the argon begins to accumulate from a baseline of zero. Because of this, volcanic layers such as ash or lava can be dated with precision. Fossils themselves are not dated with this method, but when layers above and below them are dated, we know the age of the fossil that lies between them.

Many early hominin sites are studied with potassium-based methods. Much more recently, at sites younger than 40,000 years old, the carbon-dating method is used. The death of an organism essentially starts a clock in which the radioactive carbon-14 isotope begins to decay within the fossil, and it makes up less and less of the total carbon through time.

Where radio-isotopic methods cannot be applied to fossil sites, other methods are useful. The most common is to use the types of animals that are discovered to place the site at the most logical time period, based on the dates known for those animals at sites with good calibration. This is called biostratigraphy or biochronology, and the date for *Sahelanthropus* was determined this way.

The 'Sidi Hakoma tuff' is a volcanic ash layer in Ethiopia from an eruption 3.4 million years ago. In some places these layers reach incredible thicknesses of 5–10 m (16–33 ft), because ash was piled up in a river channel.

The motion of the river current is immortalized in the rock.

Ardipithecus ramidus and *Ardipithecus kadabba*

The *Ardipithecus* genus is a group that lived between 5.8 and 4.3 million years ago. In 1994, the first *Ardipithecus* fossils were announced by a team, led by Tim White, working in the Middle Awash region of Ethiopia. Since then, similar fossils have been found at other sites in Ethiopia. As with *Sahelanthropus* and *Orrorin*, we see a shift in the anatomy of teeth that indicates a close relationship to hominins. There is also a likelihood of bipedal behaviour, but the discoverers are cautious in interpreting the skeletons. *Ardipithecus* uses the word 'ardi', meaning 'ground' in the Afar language of people who live in the area where these fossils were found. The species names refer to the view of this group as an ancestor for later hominins – in Afar, 'ramid' means 'root' and 'kadabba' means a family ancestor.

Ethiopia

The Gona site, one of the places in Ethiopia where *Ardipithecus* fossils are found.

Yet another biped?

From initial reports, the most definitive evidence for calling *Ardipithecus* a hominin is the size and shape of the teeth. Although we would see *Ardipithecus* canines as large and 'ape-like', they have begun to function more like those found in humans. The first premolar does not hone the upper canine when teeth come together. Instead, as in humans, the canines wear on their tips rather than the back edge.

Sahelanthropus and *Orrorin* illustrate the key debates about early bipedalism. Can bipedalism be determined from the skull? Could the earliest hominins be as efficient at bipedalism as humans are today? If the position and orientation of the foramen magnum provides reliable evidence (see p. 12), then *Ardipithecus* – like *Sahelanthropus* – can be reconstructed as a biped, based on parts of the base of a skull.

More evidence comes from the shape of joint surfaces in a toe bone found with the fossils assigned to the early species of *Ardipithecus*: *A. kadabba*. It shows that *Ardipithecus* was able, in some ways, to move its feet like humans. Because of our posture, human feet have to flex upwards to a greater extent than in chimpanzees, and this affects the joints between toe bones. However, the curved shape of the toe bones of *Ardipithecus* is more similar to that of apes. As we discuss again later, a nagging question when thinking about bipedalism is how to interpret the mix of human and ape features in the skeletons of early hominins.

Tiny clues

Early hominins can be recognised even from small pieces of bone if they contain the right information. This piece contains a first premolar tooth, which shows the initial signs of human anatomy in *Ardipithecus ramidus*.

Early diets

When we think about how early hominins lived, we want to know whether modern apes provide a good picture or whether our earliest ancestors were different from living species. We also compare them with what came later to see whether the hominins had a common set of behaviours.

Diet can be reconstructed from tooth anatomy, thickness of tooth enamel (the hard outer layer of a tooth), how the tooth is worn down from chewing, and in some cases the chemical make-up of teeth and bones. With *Ardipithecus*, a key feature is relatively thin tooth enamel compared to later hominins. This suggests that

Ardipithecus did not eat the hard foods that were more common in *Australopithecus* diets later in time. *Ardipithecus* probably ate many of the foods that chimpanzees eat today, especially fruit but also a variety of other foods (e.g. chimpanzees eat termites and ants and even hunt small monkeys). In its diet, *Ardipithecus* was probably more similar to *Sahelanthropus* than to *Orrorin*, which had thicker tooth enamel.

Cousins

Chimpanzees are our closest living relatives; we share 98 per cent of our DNA with them.

Are hominins 'apes'?

The apes are our closest living relatives. The bigger species – chimpanzees, gorillas, and orangutans – are referred to as great apes. Within the great apes, chimpanzees are most closely related to humans, and gorillas slightly more distant from us. Darwin recognised the close relationship of humans to African apes and predicted that the earliest human fossils would be found in Africa. He was right.

Because great apes are closely related to us, they are usually classified within the human family, Hominidae. This book focuses on a narrower group, Hominini, which means specifically the fossil species that are more closely related to us than chimpanzees are. (In other places, you may see the name 'hominid' used in the same narrow sense that this book uses 'hominin' – the difference is just in the names and reflects an older taxonomy in which apes were excluded from the human family.)

It is easy to make the mistake of thinking that the apes are a uniform group, when in reality great differences exist among them. Chimpanzees are fascinating to us because of their complex social relationships and their tendency to show many 'human' behaviours. Gorillas are much bigger in size, and they eat leaves more frequently than the other apes, who mostly eat fruit. Orangutans and gibbons (gibbons are so-called 'lesser apes') each have their own unique ways of moving around through trees, and have exceptionally long arms.

Despite differences like these, apes share a variety of features – for example, all apes lack tails, and so do humans. Apes also have relatively larger brains than most other primates. They have projecting faces and long canine teeth as well, but in these respects they are like many other primates.

How, then, do we think of the earliest hominins – the earliest human ancestors? Early hominins are often described as 'bipedal apes' because they were ape-like in the general appearance of their skulls and their relative brain size. At some point in human evolution, we cross a threshold where we begin to think of our ancestors as more like us, showing more of the distinctive features that arise during human evolution. It is worth keeping in mind that this is a subjective decision – what is ape, and what is human? There is no definite line between these categories, and as we go through the variety of hominin species, the question becomes more and more difficult to answer.

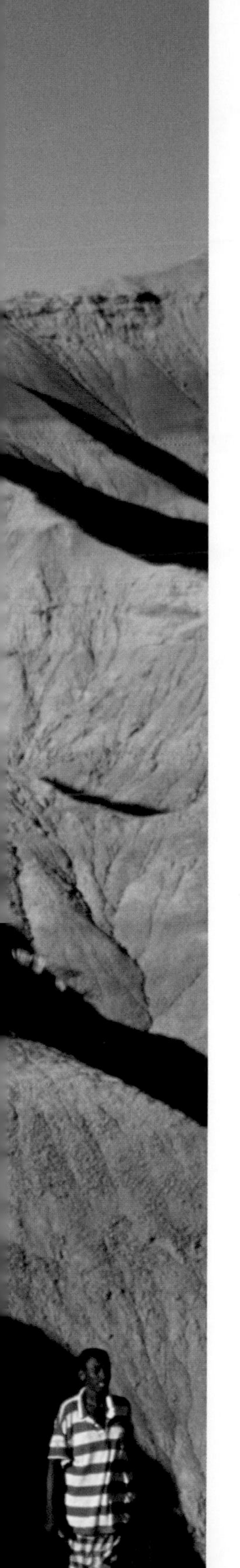

AUSTRALOPITHECUS

The discovery of *Australopithecus* in the 1920s was a key event in understanding human evolution. The full story is told in the section on *Australopithecus africanus* (see pp. 36–43). *Australopithecus* species were small-brained, bipedal human ancestors. They show us that the defining hominin characteristic is bipedal walking rather than large brains or toolmaking. As well as being bipedal, *Australopithecus* species had thicker tooth enamel than the great apes, and modified canines and premolars reflecting a change in the function of these teeth. They are thus 'hominin' in all respects. From initial comparisons, it seems that these species also behaved differently from the earliest hominins described in the previous section, so *Australopithecus* ushers in a new phase of human evolution.

Layers upon layers

The site of Hadar in Ethiopia records nearly 500,000 years in the life of *Australopithecus afarensis*. Field workers from the surrounding Afar region, such as Hamadu Mohamed Ware, play a crucial role in finding the fossil evidence.

Australopithecus afarensis

If you were to go back just over 3 million years in time and walk through the woodlands and floodplains of northeastern Ethiopia, you would see a vast array of wildlife that included a familiar creature, a hominin called *Australopithecus afarensis*. This species is familiar to us now because one of its members happened to die near a stream that quickly buried the skeleton. After millions of years of further burial and then erosion of the land surface, she was uncovered again and exposed to the elements. In 1974, Donald Johanson and Tom Gray, part of a team of researchers who had discovered the fossil abundance of the Hadar region of Ethiopia a few years before, came across first a fragment of elbow joint, then part of the skull, and eventually much of the skeleton we have come to call Lucy.

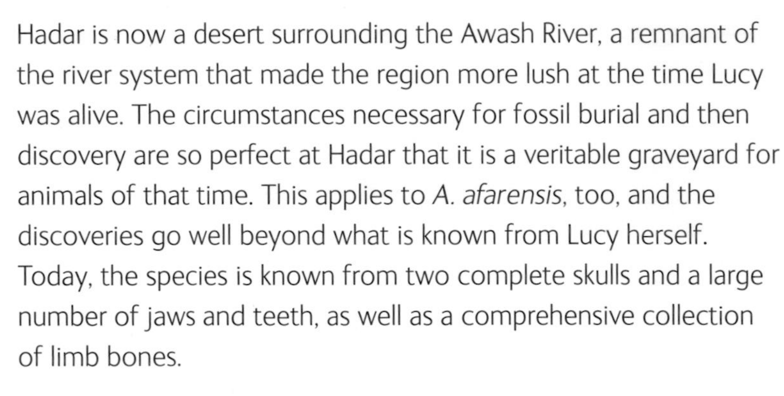

Hadar is now a desert surrounding the Awash River, a remnant of the river system that made the region more lush at the time Lucy was alive. The circumstances necessary for fossil burial and then discovery are so perfect at Hadar that it is a veritable graveyard for animals of that time. This applies to *A. afarensis*, too, and the discoveries go well beyond what is known from Lucy herself. Today, the species is known from two complete skulls and a large number of jaws and teeth, as well as a comprehensive collection of limb bones.

By looking at the shape and structure of the remains, especially the limb bones and hips, scientists can work out how hominins began to walk on two legs. It is crucial to have partial or complete skeletons to tie together all of the more fragmentary pieces that are found. This is what makes Lucy so special. When she and the other *A. afarensis* specimens were described, it transformed our understanding of human evolution in general and human walking in particular.

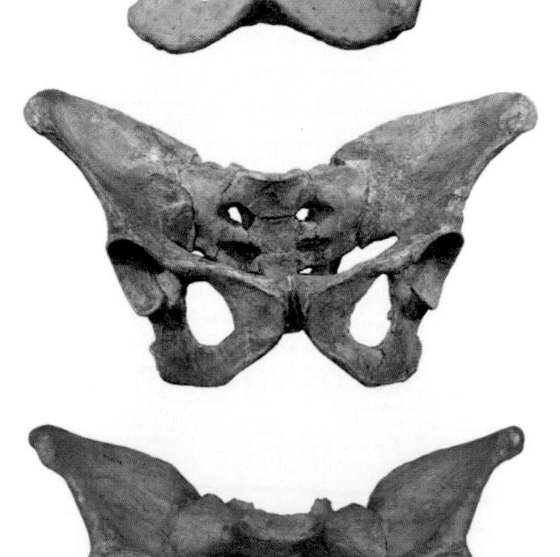

Hips

Humans (bottom) have short, broad hips compared with apes (top), and Lucy has a similar shape.

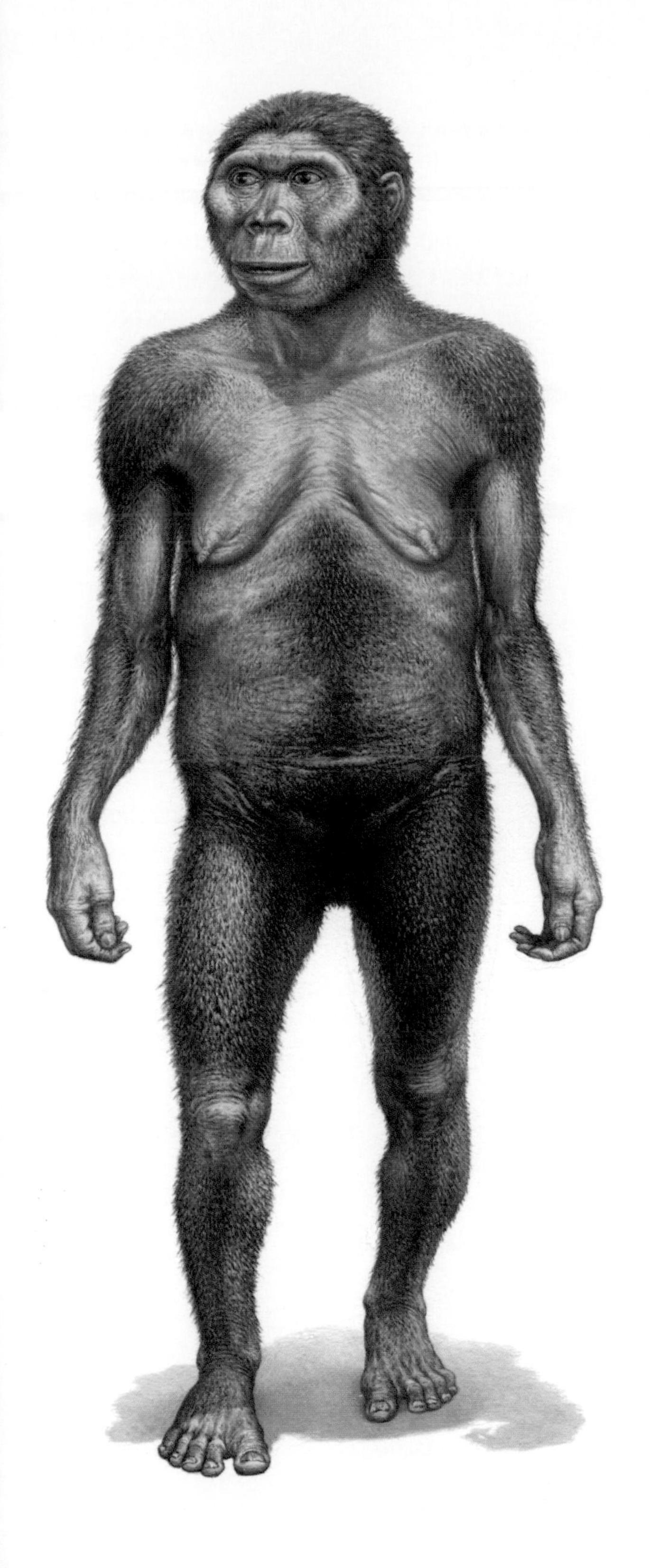

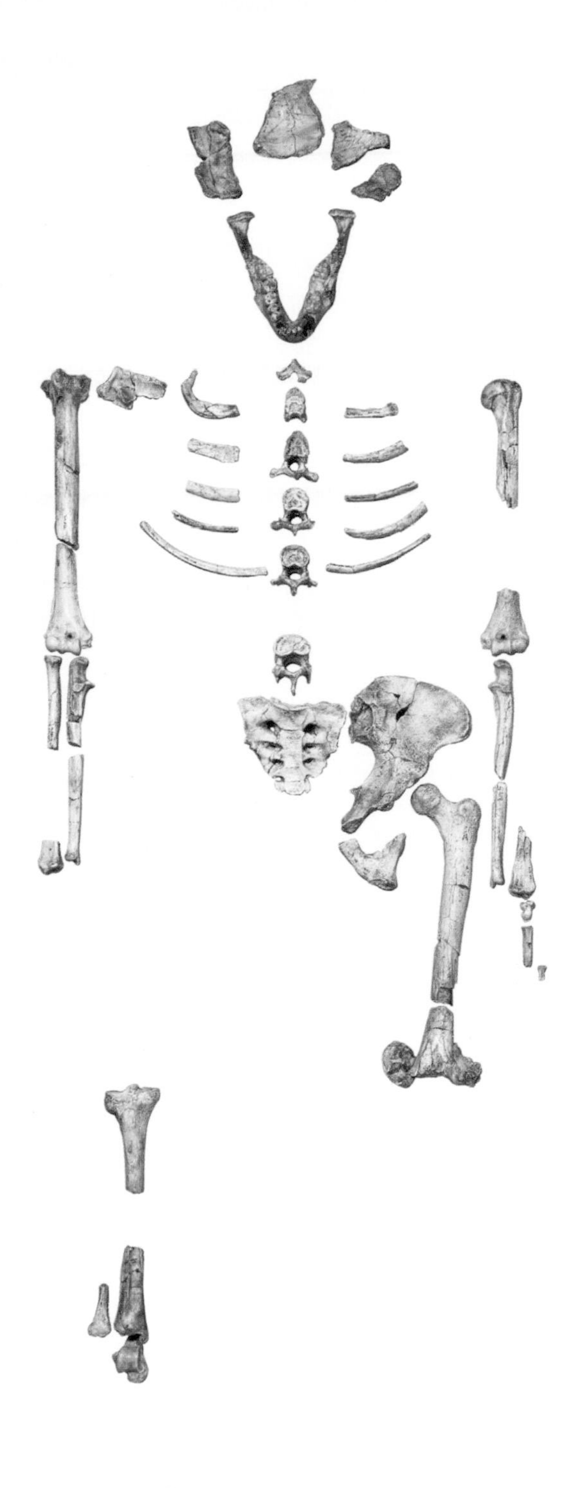

Lucy

Lucy is deservedly famous, as her discovery revolutionised ways of thinking about early hominins, especially the way they walked.

Middle aged

When named in 1978, *A. afarensis* was the oldest known hominin species. Many more discoveries have been made since then, and *A. afarensis* is no longer the oldest hominin – in fact this species now sits approximately in the middle of human evolutionary history. At 3.0 to 3.6 million years old, *A. afarensis* comes long after species that may be the first hominins (6 to 7 million years ago) and before the origin of our own genus of hominins, the genus *Homo*, at approximately 2–2.5 million years ago.

What's in a name?

Australopithecus means 'southern ape'. The name was coined by Raymond Dart for *Australopithecus africanus* in 1924, so since then species that are broadly similar to *A. africanus* have been put into this group. When *A. afarensis* was discovered, scientists decided it was a different species from *A. africanus* but that it still belonged to *Australopithecus*.

Afarensis means 'of the Afar region', the region of northeastern Ethiopia where the largest sample of *A. afarensis* fossils has been found. *Australopithecus afarensis* fossils have also been discovered

Fossil sites

Major early hominin sites in eastern and central Africa. These sites contain fossils of the earliest hominins as well as Australopithecus and Paranthropus species.

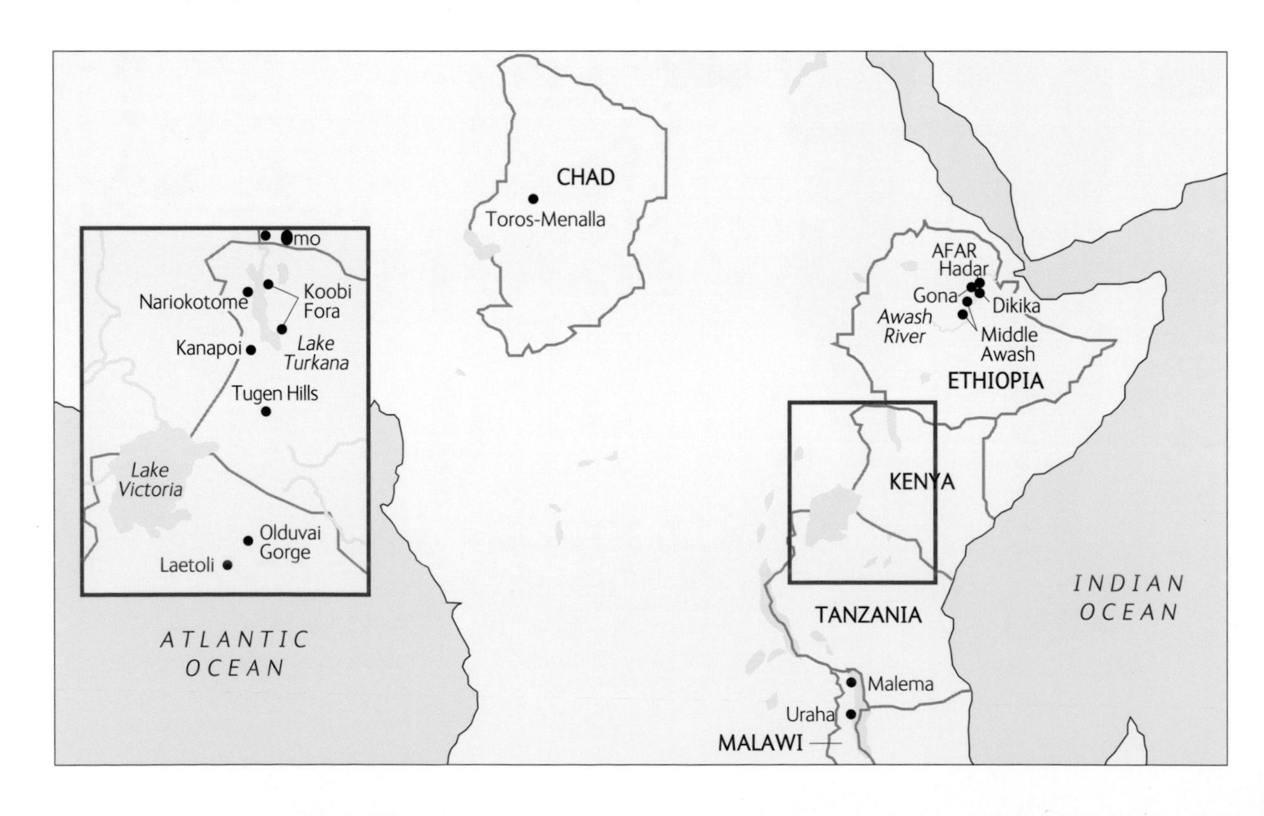

Footprints

The Laetoli trackway in Tanzania is a unique series of footprints made by three hominins walking across ash from a recent volcanic eruption at about 3.5 million years ago.

in Tanzania (at a site called Laetoli), as well as in Kenya and other sites in Ethiopia. Somewhat further away, a few specimens from Chad may belong to *A. afarensis* as well (see box on *Australopithecus bahrelghazali*, p. 34).

First steps

Lucy is not the only extraordinary find associated with *A. afarensis*. The Laetoli site in Tanzania is home to a footprint trail 3.5 million years old that is probably a trackway of *A. afarensis* individuals, because of teeth and jaws found at the same site which resemble those from Hadar in Ethiopia. The footprints, in combination with the information from the Lucy skeleton and other specimens, illustrate in dramatic fashion that hominins at this time were bipedal.

What is it about the skeleton that indicates bipedalism? As one might imagine, the hips and legs are the key evidence of walking upright. Lucy's pelvis is shorter and broader than it is in apes, and it is similar to humans in that respect. The knee joint is angled, as in humans, so that the bottom of each thighbone (the femur) is directed towards the middle of the body. This means that, when each step is taken, the centre of mass passes more directly over the lower leg. Foot bones, in combination with the footprints, show that the heel bone was pronounced, and the foot may have formed a supporting arch in the same way that human feet do.

Habitats

The habitat of *Australopithecus afarensis* was probably a mix of wooded areas along with more open territory.

Studies of *A. afarensis* have often focused on bipedalism, mainly because the sample of bones from the skeleton is more complete than for most other early hominins. Everyone agrees that *A. afarensis* was indeed bipedal, as opposed to quadrupedal (like most monkeys) or suspensory (like orangutans and gibbons). *Australopithecus afarensis* did not run along the ground on all fours, and it did not have extremely long arms for climbing and swinging. There is also no evidence that it was a knuckle-walker, the particular way that chimpanzees and gorillas walk on all fours by using the knuckles of their hands instead of their palms.

When on the ground, *A. afarensis* walked upright, not exactly with the same mechanics as humans today, but in a way that was far more similar to us than to modern apes. Differences in the exact pattern of walking are suggested by details of the pelvis and lower limb. For example, the blades of Lucy's hip bones flare out further to the side than they do in humans, and they do not curve around to the front as much. Muscles would have been in a slightly different position, leading to a different way of walking.

Ground or tree-dweller?

We can reconstruct how we think the species behaved from the skeleton. But there is some controversy over how much time *A. afarensis* spent on the ground, as opposed to in the trees. Arboreal behaviour is a major part of most primates' lives, and even primates that appear highly adapted for quadrupedalism on the ground retain general features of the skeleton that allow them to climb trees and obtain food or take refuge there.

Australopithecus afarensis was a chimpanzee-sized animal, smaller than modern humans, and many aspects of its skeleton fit with the idea that this species still used trees as a major food resource and/or refuge for safety and sleeping. In other words, it was bipedal when on the ground but was able to climb trees more effectively than modern humans. Its relatively long and powerful arms would be used for hauling itself up, and its relatively long foot, and the curved bones of its toes and fingers, would have helped in grasping branches.

How they lived

Despite being bipedal, species such as *A. afarensis* are often referred to as 'ape-like', with relatively small brains compared to modern humans. Also, when we look at the projecting face and large jaw of *A. afarensis*, we see essentially the face of an ape, but that of an upright-walking ape.

Although it had an ape-like appearance, *A. afarensis* represents a way of life that helps differentiate early hominins from their ancestors. Early *Australopithecus* species were 'megadont', meaning that their teeth and particularly their back teeth were enlarged compared to chimpanzees. The back teeth do not show the prominent crests seen in gorillas, which also have large teeth but use them to slice the leaves that form a big part of the gorilla diet. The teeth of *A. afarensis* show that it was adapted to a diet that included small, hard items such as hard seeds or nuts, which could be cracked or ground up by relatively flat teeth with a thick enamel surface. There is no doubt that much of the *A. afarensis*

Changing teeth

A set of upper teeth from *Australopithecus afarensis*, showing a more oval shape to the tooth row than seen in chimpanzees or gorillas, with smaller canines.

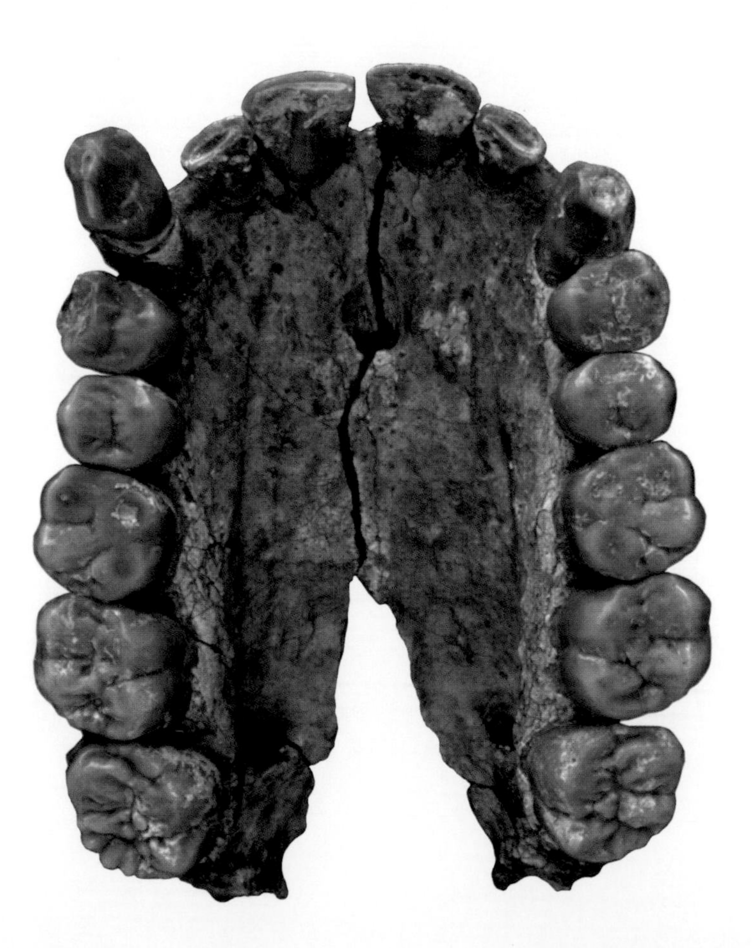

diet was fruit, but the shape of the jaws and teeth suggest that, when necessary, *A. afarensis* was able to consume harder foods than, for example, chimpanzees.

The same could be said for other *Australopithecus* species, described in other sections here. What makes *A. afarensis* different? Amongst the range of *Australopithecus* species, *A. afarensis* had a more 'traditional' primate diet, with an emphasis on small fruits and other foods such as young leaves that were easy to chew. Microscopic wear on the teeth – basically, the marks made by food items when they are chewed – shows that *A. afarensis* had a wide-ranging diet, and hard foods were probably something it ate as an important part of its diet, perhaps more during lean seasons. In other words, *A. afarensis* was not specialised for eating one type of food.

A baby girl

New fossils of *A. afarensis* continue to answer some questions as well as raise others. In 2006, after five years of intensive cleaning and preparation, a team led by Zeresenay Alemseged announced another skeleton of *A. afarensis*, but this one was of a young child, roughly 3 years old. It is from a site called Dikika, which lies across the river from Hadar. Much of the Dikika skeleton is present, including some rarely preserved elements such as the shoulder blade (the scapula), and only the second hyoid bone from the entire hominin fossil record. The hyoid bone is in the neck below the base of the tongue, and it gives information about the function of the voice box (larynx). Both of these structures are relatively ape-like in the baby skeleton, but even at its young age many of the typical features of *A. afarensis* were present.

How do they know it is a girl? Although it is impossible to be sure, this conclusion comes from the size of the adult canine. Because she was so young, she still had her baby teeth in, but the adult canines were forming inside the jaws. In apes, canines in males and females are very different in size but in humans they are much more similar. *Australopithecus afarensis* still showed some variation in canine size between the sexes, and the Dikika baby's adult canines are at the low end of the species range of size. It was probably a female.

Baby skeleton

Nicknamed 'Selam', this child is 3.3 million years old and the most complete juvenile skeleton of an early hominin.

Crucial evidence

Why was the discovery of *A. afarensis* so dramatic? Partly for the quality of the fossils, and partly because it held the title of 'oldest hominin' at the time. Also, the discovery of *A. afarensis* began a long debate about evolutionary relationships that has only become more complex with further discoveries. It was announced as the earliest human ancestor and a common ancestor of all later hominins. This in itself was uncontroversial, but further study of *A. afarensis* led to changing views of other species, particularly *A. africanus*, which until that point had been regarded as the early trunk species of the human family tree. These changing views raise difficult questions about how to use information from the skeleton to shed light on evolutionary relationships. Today, *A. afarensis* is still discussed as an ancestor to humans, but whether or not that was literally the case, the large sample of bones and teeth gives us crucial evidence of what hominin life was like after bipedal walking evolved but before stone tools transformed human evolution. It serves as a point of comparison for all hominins, both earlier and later species.

Australopithecus anamensis

Of the many places that have early hominin fossils, the Lake Turkana region has perhaps the greatest treasure trove. Largely situated in northwestern Kenya but stretching into Ethiopia as well, Lake Turkana is surrounded by fossil sites of all ages relevant to human evolution. One of these sites gives us the first glimpse of *Australopithecus*. Lucy and the other individuals of *Australopithecus afarensis* may be the best known of this group, but the earliest signs of *Australopithecus* belong to a different species – *Australopithecus anamensis*. The second name is derived from the Turkana word 'anam', meaning 'lake'.

Australopithecus anamensis fossils are known from two sites near Lake Turkana – Kanapoi and Allia Bay – and range in age from 3.8 to 4.2 million years ago. This age is right between the earlier *Ardipithecus* fossils and the later, well-known *A. afarensis* sample. Such a pattern leads some to conclude that human evolution at this time was a single evolving lineage.

Fossil hunters

Prospecting for fossils at the site of Kanapoi in Kenya.

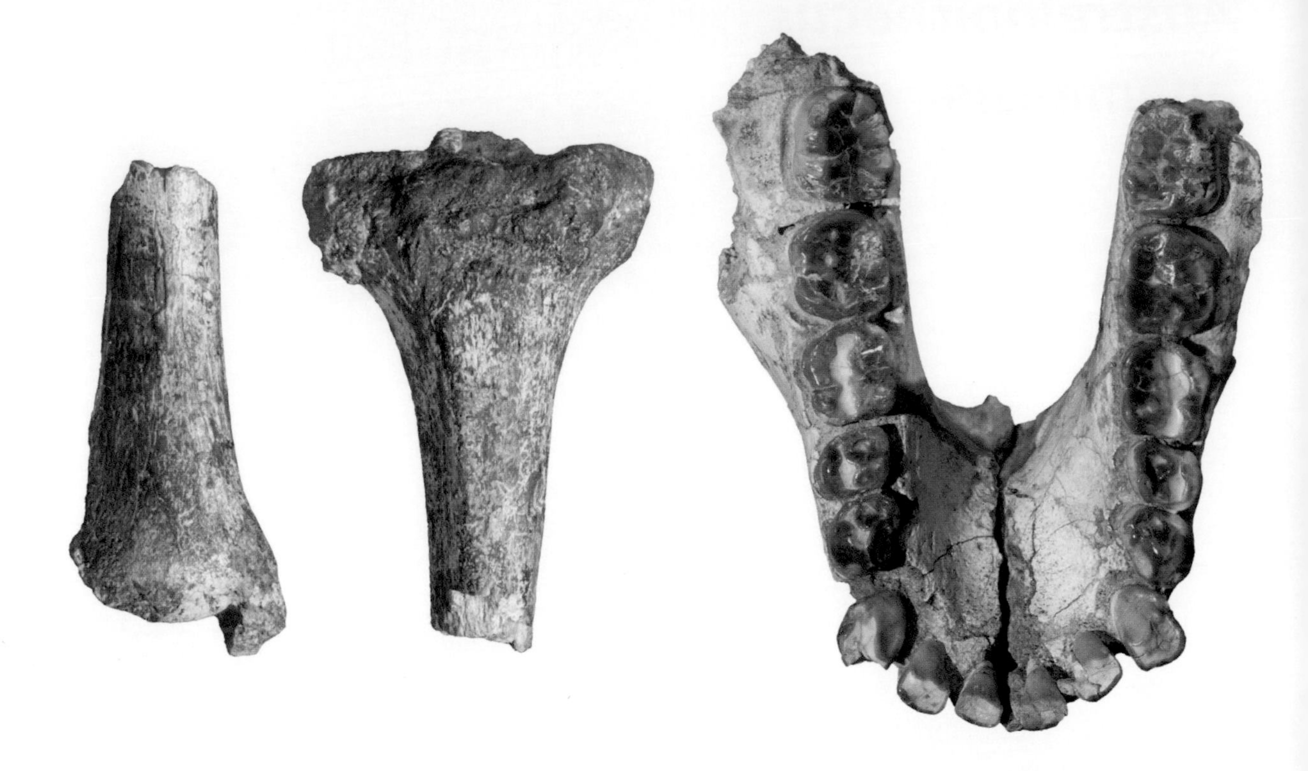

Bipedal ape

Australopithecus anamensis had a human-like lower leg (tibia), but its jaws have parallel tooth rows, more like the apes than humans.

The sample of *A. anamensis* fossils is small but contains some key evidence. Perhaps the most impressive element is a tibia (shin bone) that looks remarkably like a human tibia. The thick upper part of the tibia is robust, because of walking on two legs and transferring all weight through the legs – this is a clear difference from tibias of apes, which are less built up close to the knee. Also, the lower part of the *A. anamensis* tibia shows an ankle joint with human-like orientation. As shown in more detail in the section on *A. afarensis* (pp. 22–30), understanding bipedalism is a complex task. *Australopithecus anamensis* probably did not walk exactly like we do, but it appears that this species had a bipedal posture.

Hominins in their habitats

Australopithecus anamensis fossils are found in geological deposits that indicate a variety of settings, ranging from open grasslands to more wooded areas. Does this mean they lived in all of these habitats? Because we want to know how well they could adapt to different environments, this is an important question. Unfortunately, it is difficult to pinpoint the relationship of fossils to specific habitats. The fossil record is the result of many processes, and fossils may end up in places that don't represent where the individuals lived. We can say that there was some open country in the region where *A. anamensis* lived, but we don't know whether these hominins spent much of their time out in these open grasslands, or whether they preferred more wooded areas.

Resolving this question will help explain behaviours such as bipedalism – it is just as important to know the ecological setting as it is to know when bipedalism evolved. Lingering in many stories of human evolution is the 'savanna hypothesis', an idea that ties bipedalism and other human behaviours to more open environments. In contrast, African apes today live mostly in forested and wooded areas, and as far as we know their immediate ancestors did as well.

Four million years ago

This is how *Australopithecus anamensis* may have looked.

Australopithecus bahrelghazali

The *Sahelanthropus* skull found in 2002 (see pp. 12–13) was not the first hominin to be found in Chad. That honour belongs to a jawbone discovered in the early 1990s by the same team that would later find *Sahelanthropus*. Michel Brunet's group classified it as a new species – *Australopithecus bahrelghazali* – to recognise that it has some unusual characteristics and is geographically far away from other hominins at the time. At 3.5 million years before present, this specimen is in the same time range as the much more well-known species *A. afarensis* in eastern Africa. Because of the limited information, and the subsequent discovery of only a few other tooth specimens, many researchers regard this specimen as part of *A. afarensis*, in which case the name *A. bahrelghazali* is unnecessary.

Whatever its classification, the find is a crucial reminder that our focus on eastern and southern Africa in early human evolution comes from geological circumstance, not because hominins lived only in those places. Discoveries from places like Chad help to show the true geographical range of human evolution, and it is possible that future discoveries will reveal hominins across much of Africa. To put this in the proper context, remember that we have no significant fossil record for our closest relatives, chimpanzees and gorillas. This, of course, does not mean that they did not live in Africa in the past! Sites that provide a window into the past of west and central African forest regions have simply not been found.

The first discoveries of *Ardipithecus*, one of the earliest hominins (see pp. 16–19), played a role in rejecting simple versions of the savanna hypothesis, mainly because of the observation that some early hominin habitats were relatively wooded. Since then, the record has become more complicated, and with *Ardipithecus*, *Sahelanthropus*, and *Orrorin*, it now seems that there was a range of diverse habitats near the places where the hominin fossils were buried. Clarifying the evidence for these environments is a central long-term goal of studying human evolution.

Diversity versus lineages

Another important question in human evolution is how many species existed in the past. The number keeps growing as new discoveries are made. Sometimes, classifying species is a matter of comparing fossils from the same time period to see whether they fit patterns of variation we expect for a single species. It is like studying apes and other primates today, using their anatomy, behaviour, and genetics to determine which groups form the cohesive populations we call species. But in the fossil record, all we have are the bones. How we recognise and separate species today helps to guide us in the use of these bones to name species in the fossil record.

When fossils are found from different time periods, the task is more complicated. One must determine whether samples of fossils from the same region but different time periods are actually related as ancestors and descendants. Good evidence now exists for thinking that *A. anamensis* is the immediate ancestor for the well-known species *A. afarensis*, which lived a few hundred thousand years later and is described in the previous section. Essentially, *A. anamensis* does not show any unusual features which would indicate that it was evolving on its own lineage. Especially in the teeth, the anatomy of *A. anamensis* is intermediate between earlier hominins and *A. afarensis*, so it fits well as an ancestor to that species.

Kenyanthropus (or *Australopithecus*) *platyops*

Until the announcement of *Kenyanthropus* in 2001, many palaeoanthropologists thought that human evolution before 3 million years ago had been relatively linear. There was ample evidence for diversity later on, but the succession of *Australopithecus* species seemed to be a single evolving lineage. *Kenyanthropus* is dated to 3.5 million years ago, similar to *Australopithecus afarensis*, and it suggests that a different species existed in the same broad region of eastern Africa. Therefore, hominin lineages must have begun to diversify earlier than we thought.

Kenyanthropus was found by Meave Leakey's team in Kenya, on the west side of Lake Turkana. It has a unique combination of anatomical features that give it the name *platyops*, meaning flat-faced. Normally a 'flat face' would imply a set of features that allow greater muscle force to be used in chewing. The flatness comes from cheekbones that are well developed and further forward on the face. An important muscle for chewing (the masseter muscle) attaches to the cheekbone. In *Kenyanthropus*, however, this particular set of features occurs without many of the other expected signs of a robust chewing apparatus. Its teeth are small, and much of the rest of the skull is 'primitive', meaning that it shares a variety of features with *A. afarensis* and/or apes in general. One of the main conclusions from discovering *Kenyanthropus* was that human evolution is a mosaic process, with different species showing unexpected combinations of anatomy.

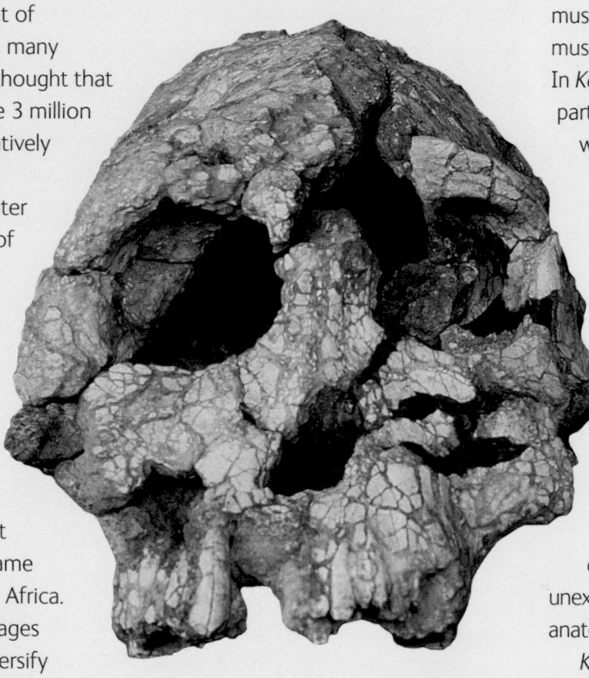

The skull of *Kenyanthropus*.

Kenyanthropus is known mainly from a single skull. The different genus name is related partly to its interesting combination of features and partly to its unknown position in the human family tree. Researchers who use a different approach to naming species may refer to it as *Australopithecus platyops*.

Australopithecus africanus

Australopithecus africanus is the defining species of its group. It was the first of the *Australopithecus* species named, and it led to the widespread recognition of small-brained bipeds as human ancestors. In other words, our ancestors stood upright long before they evolved the human behaviours related to intelligence and thought. This was a bold statement to make at the time when *A. africanus* was discovered, as until then brain size was considered the pre-eminent human feature. Now it is accepted knowledge, as the descriptions of earlier species show.

A child's skull

This revolution in understanding human history started with a tiny skull, the Taung child, announced in 1924. It was brought to Raymond Dart – an anatomist working in Johannesburg, South Africa – by miners who excavated lime from the Taung site. Dart recognised that the skull was not that of a baboon or other monkey, like the other skulls the miners brought, but instead it was clearly an ape- or human-like form. Looking more carefully, Dart concluded that the small canine teeth, the shape of the brain, and the forward position of the spinal cord (where it enters the skull through the foramen magnum) indicated that Taung was a human ancestor. It may have been a small-brained and ape-like ancestor, but the position of the spinal cord indicated it was most likely a biped.

Finding the missing link

Discovered in 1924, the Taung child from South Africa was the first specimen to be called *Australopithecus*.

Dart developed a reputation for poetic terminology, and *Australopithecus africanus* was the first species name he coined. It means simply 'southern ape of Africa', despite Dart's view that it was a human ancestor.

The evidence from Taung was not sufficient to confirm all of Dart's hypotheses, but they have since been tested from other discoveries. Depending on how you count, more fossils are assigned to *A. africanus* than to any of the other early hominin species, which makes it a fantastic test case for understanding variation. The vast majority of specimens are teeth, but early discoveries of relatively complete skulls and a partial skeleton confirmed that Dart's description of the Taung child was valid for adults as well. There are now hundreds of fossils of *A. africanus*, largely from the sites of Sterkfontein and Makapansgat in South Africa. Sterkfontein is near Johannesburg and part of the Cradle of Humankind World Heritage Site that includes other hominin sites discussed later in the book. Makapansgat is in the northern part of South Africa. In the late 1990s, a nearly complete skeleton was found at

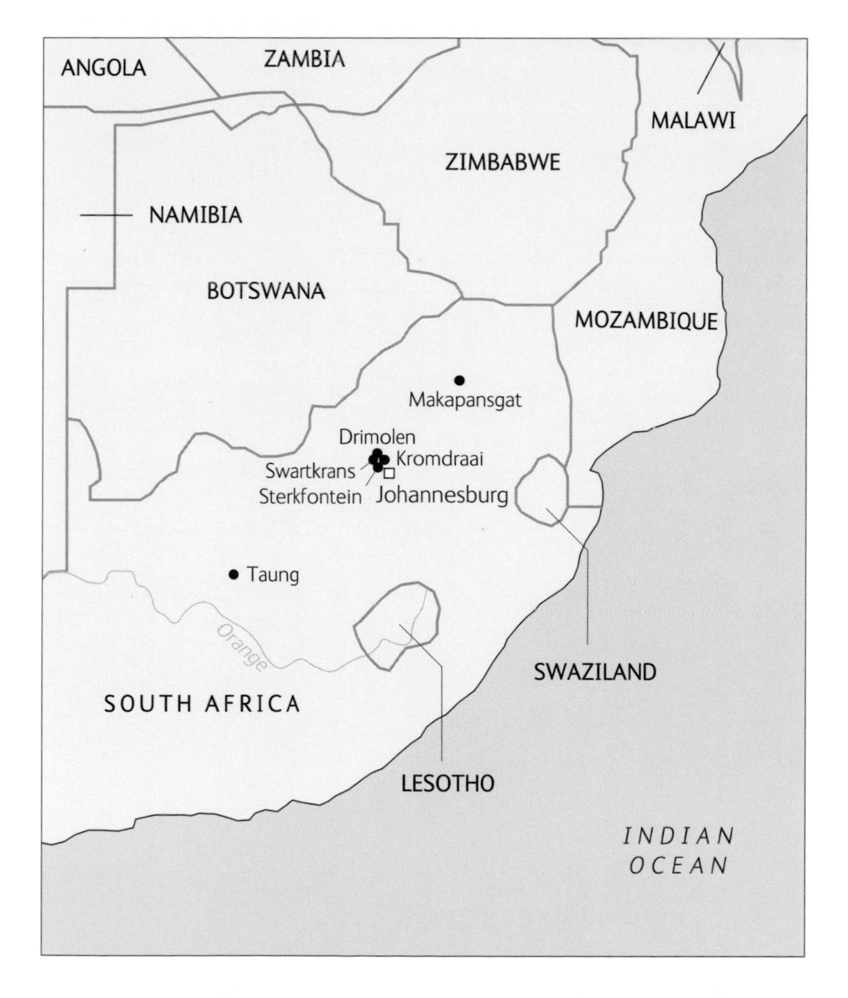

South African sites

The earliest hominins from South Africa are from diverse sites between 1 and just over 3 million years old.

Sterkfontein by Ron Clarke, in a different part of the site from the *A. africanus* fossils. The difficult excavation and cleaning of this skeleton mean that it has not been fully prepared yet, but it will no doubt tell us much about South African hominins, whether it is found to be *A. africanus* or to represent an earlier species.

Cave sites

The hominin species described so far have been found at open-air sites such as the extensive and arid badland deposits of eastern Africa and the sandy desert sites in Chad. *Australopithecus* fossils from South Africa were discovered in a completely different context – caves or remnants of caves. Caves differ from open-air sites in how the fossils get into the deposit and how the sites represent time periods. Caves tend to capture specific intervals of

Excavations

Most specimens of *Australopithecus africanus* are from the Sterkfontein site in South Africa, which first revealed hominins in the 1930s. It has now been excavated almost continuously since the late 1960s, producing hundreds of specimens.

time, when they are open for bones, stones, and sediment to enter. This material is washed or dropped in, dragged in by carnivores, or it represents animals that have fallen in and become trapped. Later cave sites are sometimes places where humans lived, but this was not the case for a species like *A. africanus*.

Cave geology can be highly irregular, in contrast to the clear horizontal layers that are often seen in open-air sites. Because cave sections can fall or shift over geological time, the general principle that younger sections lay over older sections (the geological 'law of superimposition') does not always apply. Nor are the fossils interspersed among volcanic eruption sequences, as they typically would be in eastern African sites (see pp. 24). The crucial task of dating fossils therefore becomes more complicated.

While sophisticated geochemical methods of dating are increasingly being applied to cave sites, most of our understanding of time comes from the animals themselves. Certain groups of animals are found at different times, and it is possible to develop and refine a timescale based on these cohorts by studying fossil sites with well-established dates. These dates make up a 'biochronology' of how animals have evolved and changed over time, so that when a new site is found, the animals can be compared to the patterns from elsewhere.

Keeping a lookout

Australopithecus africanus roamed among the hills near what is now Johannesburg, South Africa.

Based on such comparisons, the *A. africanus* fossils from Sterkfontein and Makapansgat are approximately 2.5 to 3.0 million years old. Whether the fossils accumulated over a long period of time or a short one is uncertain, although patterns within Sterkfontein suggest a long period of time. What is clear is that *A. africanus* fossils are found in the absence of stone tools. At least at this time in South Africa, hominins had not yet begun to fashion tools out of stone.

Ecology and diet

The main *A. africanus* sites show a mixture of habitats, from wooded areas to more open bushlands, and there were also rivers nearby. We don't know exactly which habitat would have been preferred by the hominins, or whether *A. africanus* could range successfully through the various environments.

The diet of *A. africanus* has traditionally been contrasted with species of the next section – *Paranthropus*; it did not often eat harder foods. Like *Australopithecus afarensis*, *A. africanus* probably had a broad diet but also the ability to fall back on harder foods when the opportunity or need presented itself. However, *A. africanus* did have somewhat larger chewing teeth than *A. afarensis*, and the skull morphology indicates a biomechanical rearrangement of muscles, so that force could be applied more efficiently on the back teeth.

Several years ago, researchers looked at another kind of evidence for diet. Proportions of carbon isotopes in teeth can tell us about diet because different kinds of plants process carbon in different ways. When young animals eat plants, a carbon isotope 'signature' is built into the developing teeth, and this signature is usually resistant to change during millions of years of fossilisation. A downside to this process as far as the evidence is concerned is that carnivores have a similar signature to the animals they prey on, so it can be difficult in some cases to distinguish whether hominins were eating plants or animal foods.

The carbon isotope proportions of *A. africanus* teeth suggest that its diet included a significant amount of grass, grass seeds, or other animals who themselves ate grass. But there is no

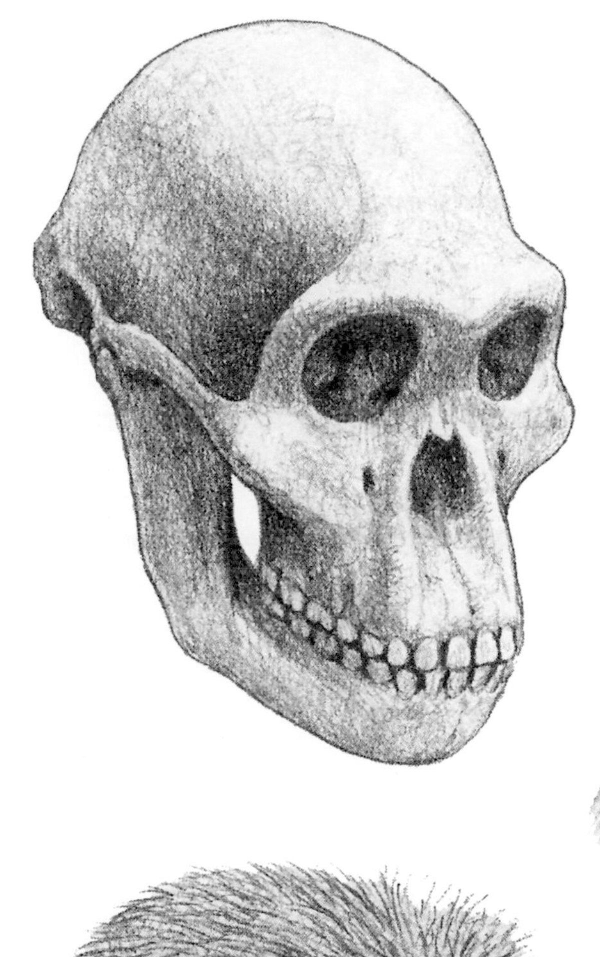

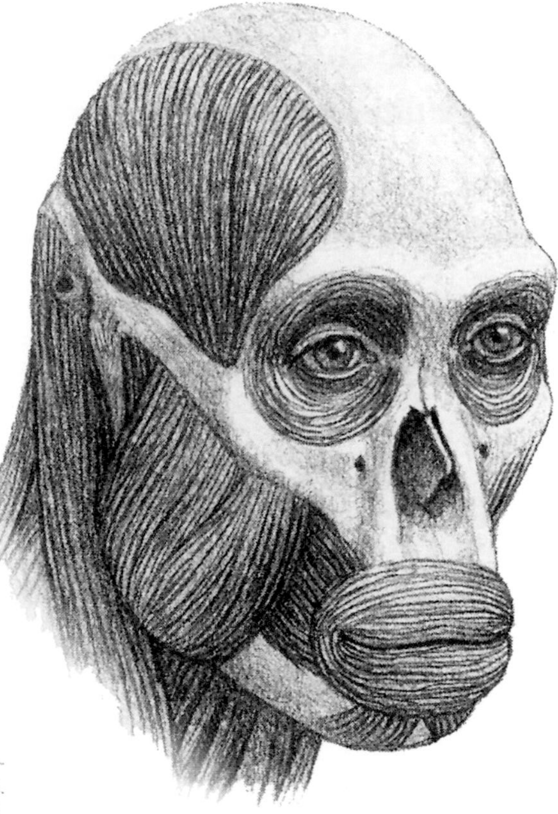

Skull shape

Skulls of *Australopithecus africanus* are an interesting mix of features shared with earlier and later hominins. They had projecting faces like earlier hominins, but cheek bones were moved forward like *Paranthropus*, and the braincase was shaped like *Homo*.

evidence in the teeth that *A. africanus* was specialised for eating grass or grass seeds. Does this mean we should choose the alternative explanation – that *A. africanus* was more carnivorous than we thought? Traditionally, only members of the genus *Homo* were considered to be meat-eaters.

Life before stone

Even if it used simple tools, *Australopithecus africanus* lived before hominins started to make recognizable stone flakes and choppers.

Chimpanzees are an example of a species that hunts animals for a small part of its food, but does not show the types of adaptations we expect for carnivorous mammals, such as teeth that can slice meat effectively. In fact, less than 5 per cent of the chimp diet is meat. A species like *A. africanus* could have been similar to chimpanzees in this respect, without us being able to see evidence of the behaviour in the bones. But if meat was a major part of the *A. africanus* diet, one would expect some other evidence, either in the shape of teeth or in the presence of stone tools. This question remains a puzzle, and work continues to try to integrate the different lines of evidence for diet.

Sex differences

Forensic anthropologists can determine the sex of a modern human skeleton with 90–95 per cent confidence if a pelvis is available, and even have an 80–90 per cent success based only on skulls alone. Can we do as well with fossil species? Because there are many *A. africanus* specimens compared to most other hominin species, we can investigate sex differences (generally called 'sexual dimorphism') with greater confidence.

Most people know that men have larger browridges than women, and deeper jaws. Such features are mentioned when we talk about 'masculine' and 'feminine' faces. In *A. africanus*, as in humans, the larger individuals tend to have more pronounced browridges. It makes sense to see this as a sex difference, since males are on average larger than females in humans and the apes (this pattern reaches an extreme in gorillas and orangutans). Because the differences are consistent with sex differences, it would be a mistake to classify large individuals and small individuals into different species. In this way, the patterns of variation we see today in humans and apes help us to understand the fossil record. And, because *A. africanus* is so well known in its skull anatomy, its variation in turn can help us to understand other hominin species.

Australopithecus garhi

Australopithecus garhi was announced in 1999 as a 'surprising' hominin find ('*garhi*' means 'surprise' in the Afar language spoken in the region where the fossils were found). The fossils are about 2.5 million years old, and the species was defined mainly by one partial skull that possesses very large teeth. A partial skeleton was found at a nearby site, but this did not have a piece of skull, so we cannot be sure that the skeleton and the skull belong to the same species (hominin species are usually defined based on skulls and teeth).

If all of these fossils do represent the same species, then the most interesting feature is the relatively long thighbone. Long legs are a human characteristic not seen in other *Australopithecus* species. The forearms of *A. garhi* are also relatively long, and some have compared the arm proportions to modern orangutans. This doesn't mean it is a fossil orangutan, but simply that it retained an ability to climb effectively. Is that pattern a contradiction with the long legs, usually associated with efficient bipedalism? Only time and further investigation will tell us.

Perhaps more intriguing than the fossils themselves were the cut-marked bones found nearby of the same age of 2.5 million years ago. 'Cut marks' are the signs of stone tools being used to carve up animal carcasses and direct evidence for hominin consumption of meat. The cut-marked bones near *A. garhi* are currently the oldest such evidence, but the research team was careful to recognise that *A. garhi* may not have been the toolmaker and user. It is very difficult to be sure about the association of particular species with the tools found at the same sites. *Australopithecus garhi* may well have more surprises in store.

The skull of *Australopithecus garhi* is broadly similar to other *Australopithecus* species but has very large teeth.

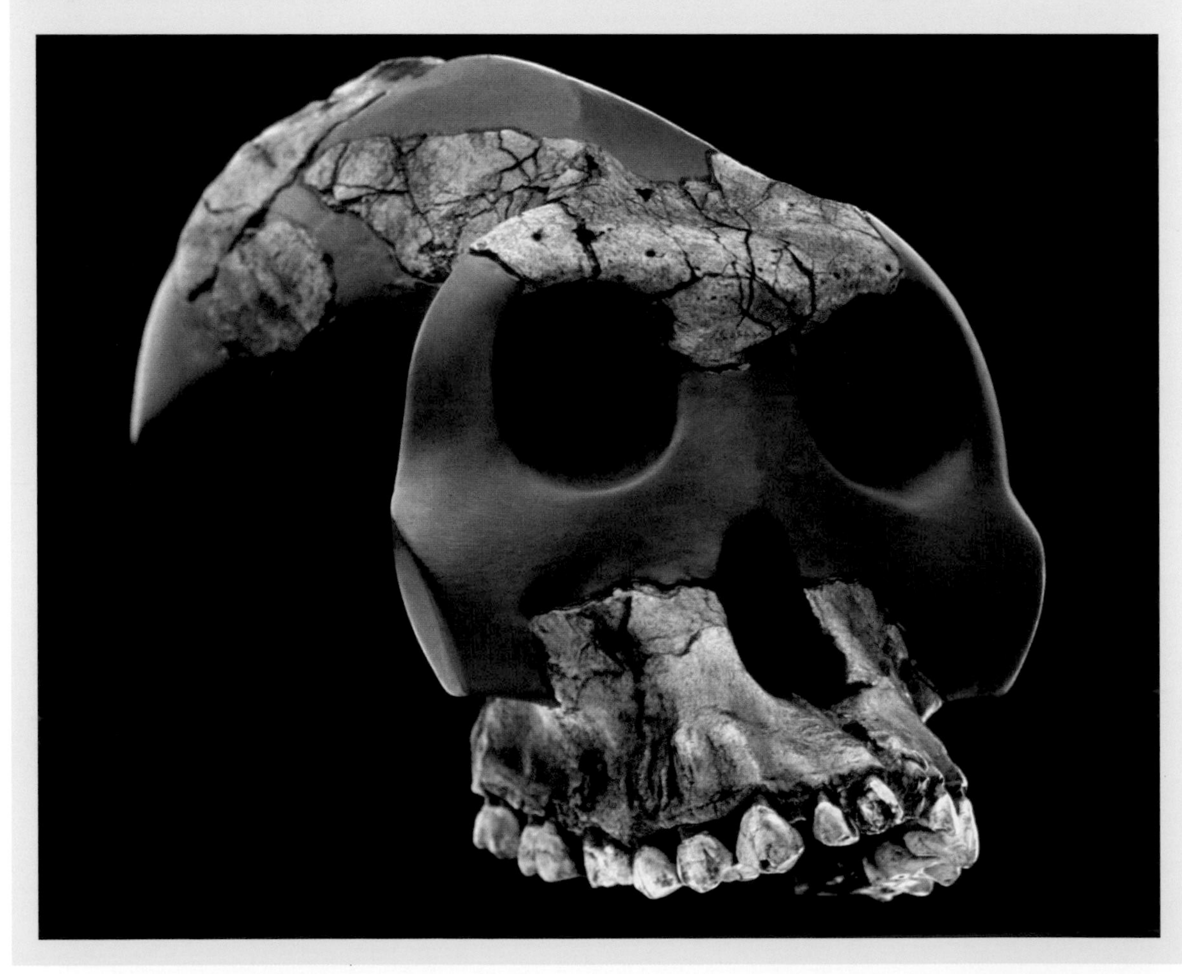

Who is related to whom?

Australopithecus africanus has long been considered a likely ancestor for our own genus *Homo*, because it shares more features with *Homo* species than do the *Australopithecus* species. But as with most things in human evolution, opinion about this has shifted over time, and today the relationship of *A. africanus* to other hominins is intensely debated. Its age suggests it could be ancestral to *Homo* and *Paranthropus* species, as it is sandwiched between *Australopithecus afarensis* and these later groups. Difficulty comes from two sources: firstly, the discoveries of other species close to 2.5 million years ago, when *Homo* and *Paranthropus* probably originated, and secondly, the anatomy used to reconstruct relationships gives different and conflicting messages. *Australopithecus africanus* has been portrayed as an ancestor to *Homo*, an ancestor to *Paranthropus*, or an isolated lineage that became extinct. It is fair to say in this case that we don't yet have the answer.

Where did *A. africanus* come from? This question is easier to answer. Based on present evidence, there are older hominins from eastern Africa than southern Africa, so we conclude that *A. africanus* arrived in South Africa as a hominin migrating or spreading south.

PARANTHROPUS

The *Paranthropus* group is one of the most fascinating branches of the human family tree, largely because they reveal to us a radically different way of being hominin. In a real sense these species were our cousins, not directly on the human line but well within the hominin group. At approximately 2.5 million years ago they diverged from our own lineage and came to be defined by an adaptation to eating hard foods like nuts, seeds and tubers. All of these species are regarded as bipedal, but we are uncertain about the details because not many skeletons are known for *Paranthropus*. The other name for this group is 'robust *Australopithecus*'. This name is used by scientists who treat them as part of the *Australopithecus* group.

Olduvai Gorge, Tanzania

Fossils found here include species of *Paranthropus* and *Homo*, as well as numerous stone tools.

Paranthropus robustus

What is the hardest thing you can chew? The answers for a human are not very impressive, as we have evolved a lifestyle that involves cooking hard or tough foods so that they are easy to chew and digest. If you were *Paranthropus robustus*, the answer would be completely different. They were able to generate massive forces to crack open hard objects, and signs of this ability make their skulls quite distinctive.

A 'dished' face

A skull of *Paranthropus robustus*, showing the prominent cheekbones that supported powerful chewing muscles.

The first evidence of their remarkable skulls came when *Paranthropus* fossils were found in South Africa in the 1930s. A doctor and palaeontologist named Robert Broom had taken up the search for hominins in South Africa, and his work would prove to be central in defining the hominin fossil record. When describing the face of *P. robustus*, Broom noted an unusual pattern – the cheekbones projected further forward than the nose. If you placed a pencil sideways across the front of the face, it rested on the cheekbones, and the nose would sit back behind the pencil. Broom called their faces 'dished' because the forward projection of the cheekbones made the face resemble a shallow dish.

This sort of appearance is partly because early hominins had relatively flat noses like you would see in a gorilla or chimpanzee. But, more importantly, the cheekbones were brought forward in *P. robustus* to put its chewing muscles in a more effective position. These bones were also very large and robust, and able to support stronger chewing muscles than seen in other hominins. Beneath them was a battery of large premolars and molars – all of the teeth that sit behind the canine and are mainly used for cracking, grinding or slicing foods.

The name *Paranthropus* means roughly 'equal to' or 'like' humans, and the name *robustus* conveys the impression of large teeth and chewing muscles that has lasted as a general description of this group. The name in this case conveys a real meaning. The species *P. robustus* is not one of our direct ancestors but is simply near us in the evolutionary tree. In many ways, this makes species such as *P. robustus* more interesting, because it shows that human

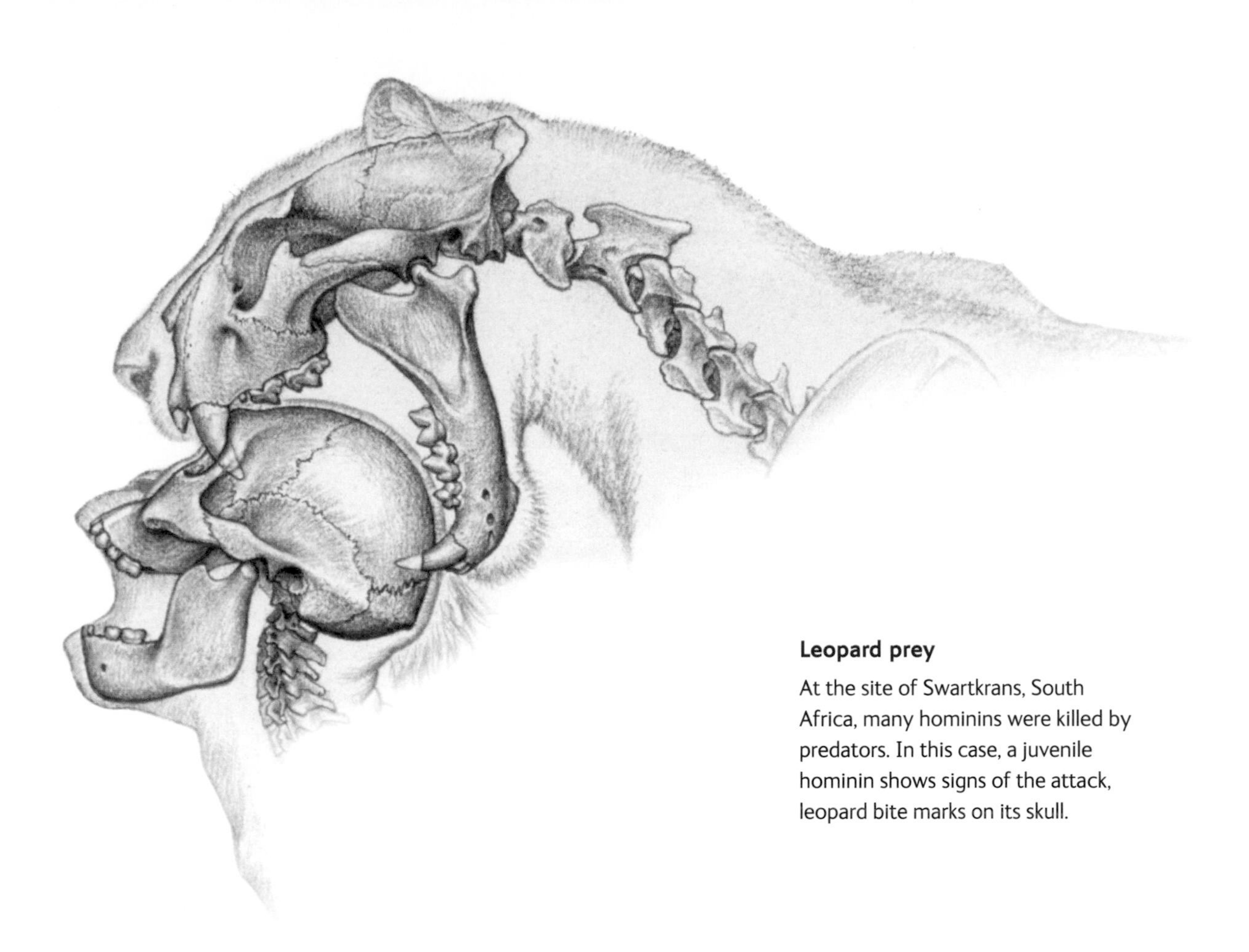

Leopard prey

At the site of Swartkrans, South Africa, many hominins were killed by predators. In this case, a juvenile hominin shows signs of the attack, leopard bite marks on its skull.

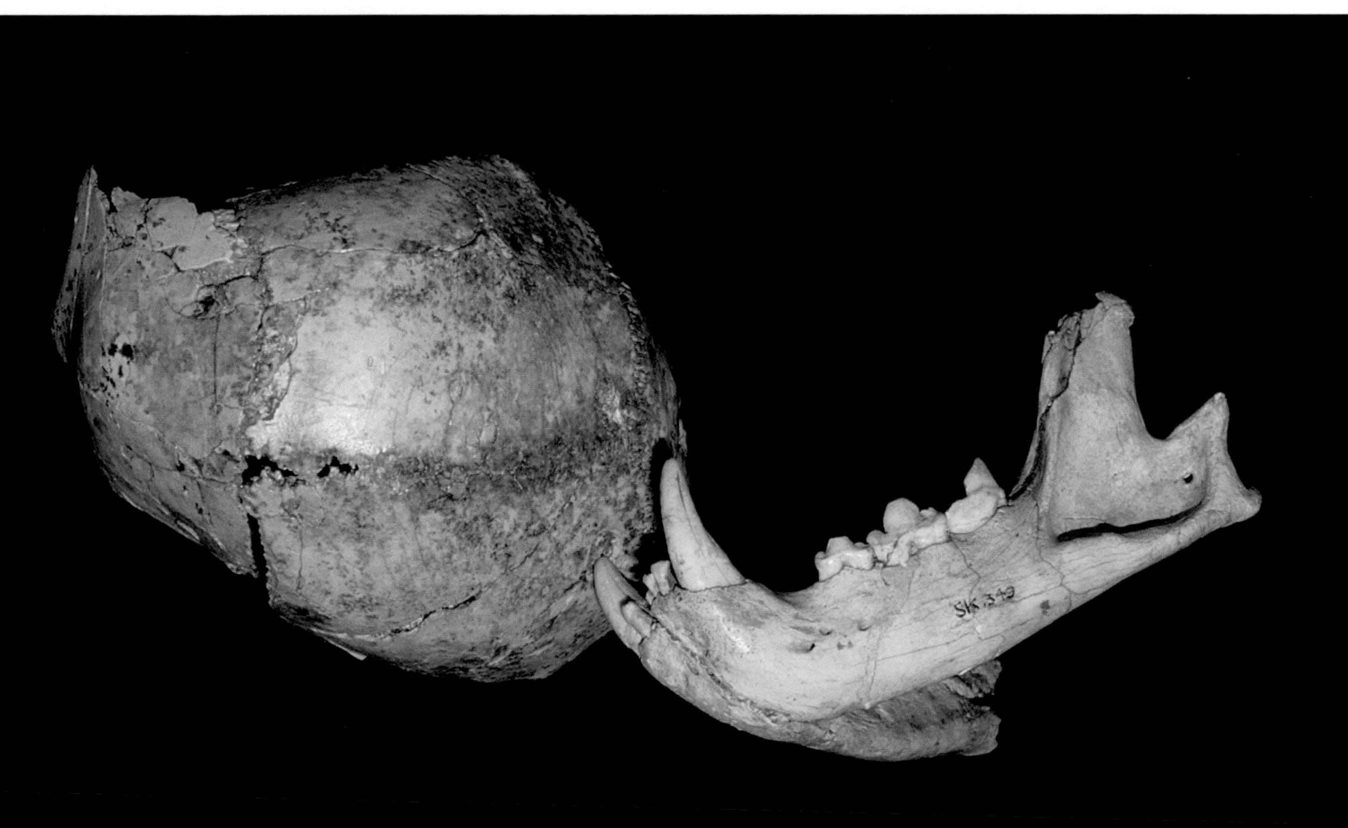

Cave sites

Like other hominin sites near Johannesburg, Swartkrans is a former cave. At various times, it was filled from above with sediment, bones, and tools, then later exposed by erosion.

evolution was not a single path towards modern humans, but instead a more complicated and diverse array of forms. There is more than one way to be a hominin.

The description is also appropriate because fossils of *P. robustus* are usually found at the same sites as our closer relatives, species of *Homo* that are described later. So, if you visited South Africa between 1.5 and 2.0 million years ago, you would see two different kinds of bipeds. One would look relatively familiar as it used tools to obtain meat and vegetable foods. The other – *P. robustus* – would have moved across the landscape consuming mostly vegetable foods and particularly the hard nuts, seeds and tubers that would have been difficult for our own ancestors to chew. *Paranthropus robustus* may also have preyed on small animals for meat, and it probably used simple bone or stone tools. There is also evidence for seasonality in its diet, which might be expected for many early hominins. But, even with a broad diet, *Paranthropus* apparently focused on different resources from *Homo* in order to survive. Both kinds of species were successful in their environments, and indeed many scientists explain their evolution in similar terms. Climates were changing in Africa at the time, producing more open habitats such as grasslands and savannas, and *P. robustus* and *Homo* were adapted to these new environments in different ways.

Paranthropus boisei

If there were a ranking of 'extreme hominins', *Paranthropus boisei* would be at the top of the list. While it fits the general description of *Paranthropus robustus*, *P. boisei* takes the same set of adaptations and skull features and exaggerates them. Most scientists view these two species as close relatives, one found in southern Africa and the other found in eastern Africa. The range of *P. boisei* stretches from Ethiopia down to Malawi.

Because of the geological differences between most eastern African sites and the southern African cave sites (see the box on p. 15), we can establish when *P. boisei* was around more accurately than for *P. robustus*. *Paranthropus boisei* lived from 2.3 million years ago until some time after 1.4 million years ago. In other words, it lived in Africa for roughly a million years of time before it went extinct. By comparison, modern humans have only been around for about 150,000 to 200,000 years.

The first hominin of Olduvai Gorge

The first specimen discovered for *P. boisei* is one of the most dramatic, as it is very well preserved and shows us most of the main aspects of the *P. boisei* skull. It was also the first major hominin find from Olduvai Gorge, the famous site in Tanzania where Mary and Louis Leakey did much of their work. The Leakeys had discovered stone tools at Olduvai, but by the late 1950s there was still no evidence of the maker of these tools. The discovery of *P. boisei* provided a potential toolmaker but perhaps not what had been expected. Instead, like *P. robustus*, it showed more evidence of a different lineage in human evolution. The Leakeys would go on to find evidence of *Homo habilis* and *H. erectus* alongside the *P. boisei* fossils. As in South Africa, it seemed that these two different groups of hominin – *Homo* and *Paranthropus* – lived in the same regions.

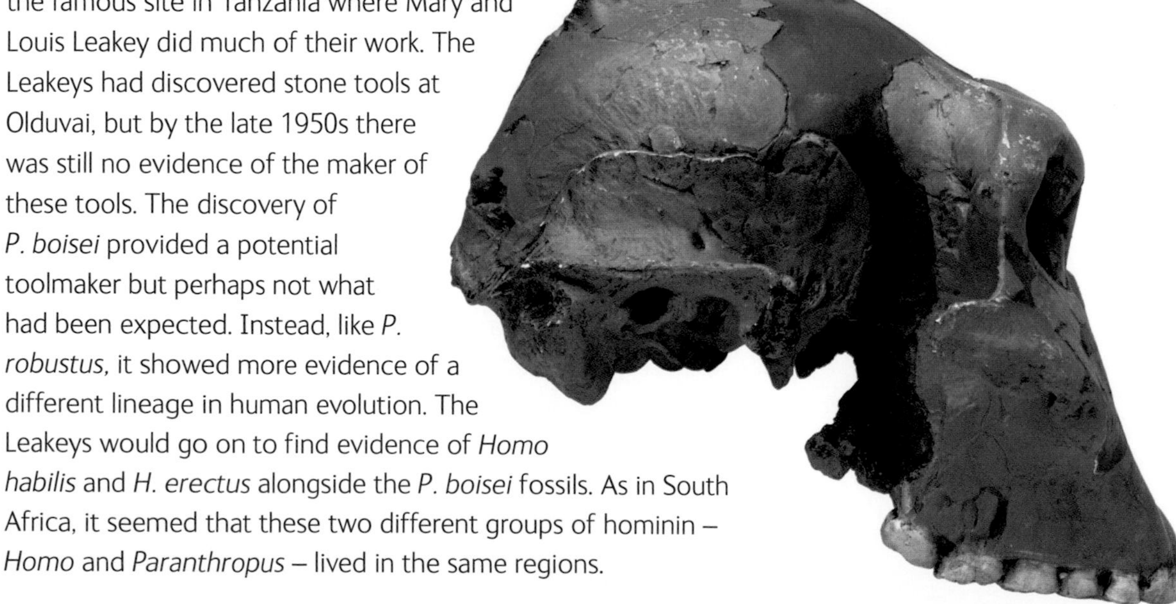

The 'Zinj' skull

The skull of *Paranthropus boisei* from Olduvai Gorge, Tanzania, nicknamed after the first scientific name it was given *(Zinjanthropus)*.

Koobi Fora, Kenya

Meave and Louise Leakey prospecting in the region of Kenya where abundant fossils of Paranthropus boisei are found.

The name *boisei* given to this species doesn't signify anything about the hominin itself. It was a tribute to the Boise Fund, the organisation that supported the work leading to the discovery of the species. Since it was discovered, *P. boisei* has become central in discussions of human evolution, as large numbers of fossils have been found. These verify the species' distinct anatomy, as seen in the 'dished' face described in the section on *P. robustus*.

My, what small canines you have

Most of us know from television programmes or visits to zoos that monkeys can be very threatening when they open their jaws to reveal large canines. This is especially true of the larger monkeys, such as baboons and mandrills. It is of course a crucial part of their life to be able to make these threats or to use the canines as weapons. In apes, such as chimpanzees and gorillas, the canines are less daunting but still relatively large. In humans, they are small and no longer serve the same function. In *P. boisei*, all of the front teeth were tiny, including the canines.

In a sense, *P. boisei* put all of its tooth-making energy into its back teeth. The difference in size between the tiny, peg-like front teeth (incisors and canines) and the enormous back teeth (premolars and molars) is extraordinary. The teeth tell us that *P. boisei* ate large quantities of small objects, placing nuts or seeds into the back of the mouth and chewing them up. This behaviour is distinct from the common way that primates eat, where front teeth are used more extensively to rip flesh off fruits or crop off leaves.

It also means that despite its radical adaptations to diet, *P. boisei* had something in common with all other hominins. The human family is defined foremost by having smaller canines than the apes.

A chewing machine

Paranthropus boisei (top) has much larger back teeth than *Australopithecus afarensis* (bottom), but its front teeth are tiny.

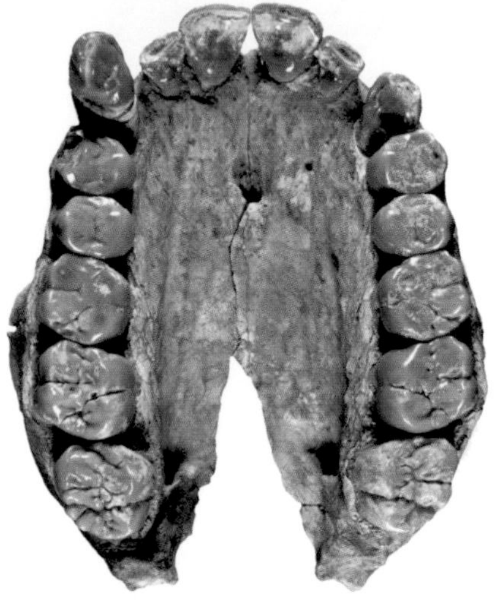

Tooth wear

We tend to think of teeth as hard surfaces, in fact the hardest surfaces in our skeleton. If you have always eaten cooked foods and gone to the dentist regularly, the surfaces of your teeth probably look generally the same as they always have, aside from cavities and the dental problems that slowly accumulate. So what would it take to wear the surface away entirely? For other primates, and for most humans in the past, this is a common thing to happen. With a less prepared diet, teeth slowly wear away as they are used to chew foods. The harder the food, the faster the tooth wears down. At some point, all of the outer layer (the hard enamel) is gone.

This process happened at an amazing rate in *P. boisei*. Despite the fact that *P. boisei* had a thicker enamel layer than any other hominin, the enamel would wear away after just a few years of chewing and grinding up foods. The best illustration of this is the first skull found for *P. boisei*. We can tell from the wisdom teeth that this individual had reached the same point of growth as a human between 15 and 20 years old. (His actual age was younger, because *Australopithecus* and *Paranthropus* species grew up faster than modern humans, but in any case he was not yet adult.) Despite being young, he had already worn away much of the tooth surfaces of the adult teeth that appeared earlier. While tooth wear was pronounced in other hominins, too, groups such as *Australopithecus* do not show this extreme pattern.

Boys and girls

One of the most intriguing features of *P. boisei* and *P. robustus* is the variation in skull size among individuals. Skull size and anatomy were very different between males and females, on a par with the differences seen in gorillas. In gorillas, males are roughly twice the body weight of females. This ratio is much greater than in humans and chimpanzees, in which males tend to be larger than females, but not dramatically so. We can't be sure about body size in *Paranthropus*, because little is known of the whole skeleton for these species. However, the skulls show a range of size and shape that allow us to determine the sex of individual specimens. For example, larger skulls have a prominent crest running along the top of the skull, but the smaller individuals do not. This crest is a side effect of hyperdeveloped chewing muscles (specifically the

Different size, same shape

As in gorillas, males of *Paranthopus boisei* were much larger than females.

temporalis muscles). The large muscles reach all the way up from the jaw to the top of the skull, and the extra crest gives them more space to attach.

If this feature helps to identify male skulls, as it does in gorillas, then the larger specimens of *P. boisei* are males, and the smaller ones are females. The degree of size difference leads some to believe that *P. boisei* and *P. robustus* had a similar social system to gorillas, in which large males tend to control groups of females and have most of the offspring while the group is maintained. We can picture small groups of *P. boisei* moving through eastern African woodlands, grasslands and floodplains, with one or two big males and a larger number of females. This is essentially the image of gorillas but in a different, more open landscape. While simplistic, this picture helps illustrate that human evolution involved a variety of different behaviours in different species.

Not just a pretty face

As in *Paranthopus robustus*, the dominant features of the *Paranthopus boisei* face are the massive cheekbones, adapted to a diet of hard foods.

Paranthropus aethiopicus

Perhaps the most evocative nickname in the human fossil record is 'The Black Skull'. That name was given to the discovery in Kenya that was announced in 1985 and essentially defines *Paranthropus aethiopicus*. The Black Skull was so called because it is literally black, but the name is also appropriate because the discovery caused much disagreement among scientists about its place in human evolution. Based only on *P. robustus* and *P. boisei*, *Paranthropus* would appear to be a relatively uniform group of species, but *P. aethiopicus* is somewhat different from the others and shows an unusual combination of features. Indeed, in the 20 years since the Black Skull was announced, scientists have not been able to reach a consensus about evolutionary relationships among early hominins. They agree about the big picture, but details about species such as *P. aethiopicus* are unresolved.

In some ways, *P. aethiopicus* shows the beginning of typical *Paranthropus* anatomy. It had large chewing muscles and some aspects of the facial appearance of other species in this group. But it also had a more projecting snout, as seen in *Australopithecus* species. The simplest way to see it is as an intermediate species. The original skull dates to 2.5 million years ago, and the few other specimens known for this species are dated to about the same time period, so perhaps this was not a long-lived species compared to others.

To complicate matters, the Black Skull shares some distinctive features with only *P. boisei*, and some scientists would include it in that species. But that would mean that the southern African *Paranthropus*, *P. robustus*, evolved on a completely separate lineage. The similarity between *P. robustus* and *P. boisei* is so pronounced that most scientists think this unlikely. It seems more likely that *P. robustus* and *P. boisei* had a common ancestor that had all of the features they share, but *P. aethiopicus* does not quite fit that description.

A mosaic of evolution

Paranthropus aethiopicus is one of those species that illustrates the pattern of 'mosaic' evolution (see also *Kenyathropus* on p. 35). This term is used to describe a pattern when features evolve independently and evolve multiple times. A broader example is the separation of human bipedalism and human brain size. We may think of humans as large-brained, bipedal primates, but we became bipeds long before brains reached anything like the size they are now. Human evolution displayed a mosaic pattern instead of a simple progression from one thing to the next. At a more detailed level, unusual discoveries like the Black Skull, which doesn't fit simple notions of evolutionary paths, also illustrates mosaic evolution.

Unfortunately, the Black Skull all by itself is a large part of what we know about *P. aethiopicus*. Most of the other specimens are fragmentary bits of teeth and jaw, and it is unclear whether the one good skull is typical of its species. *Paranthropus aethiopicus* thus illustrates another common problem in studying the fossil record – sometimes the number of specimens is too small to be sure about the details.

An unusual combination

Nicknamed 'The Black Skull', this specimen from Kenya is 2.5 million years old and the first sign of the *Paranthropus* group.

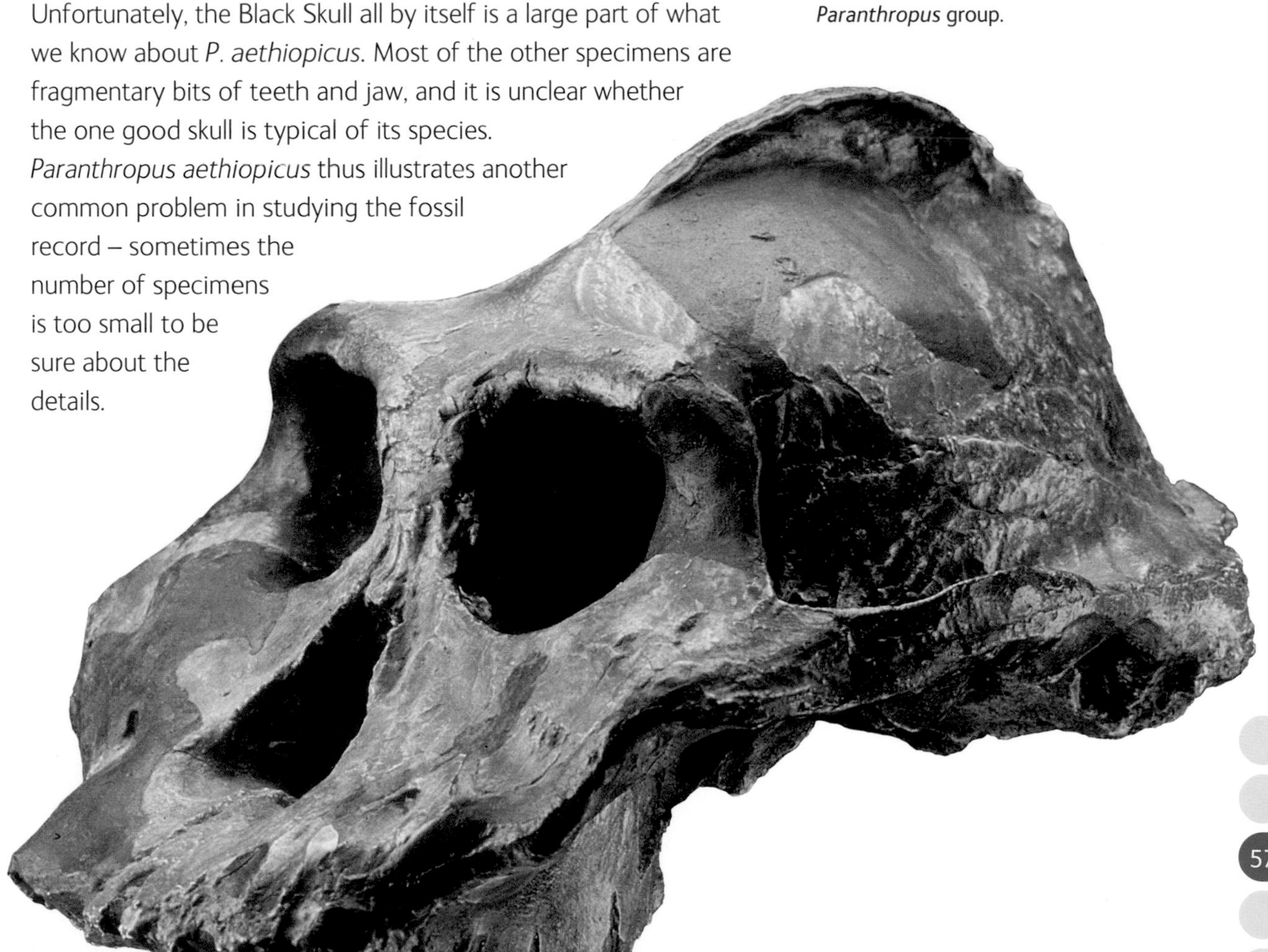

EARLY *HOMO*

Of all the names used to describe species in human evolution, *Homo* is the one that means 'human'. It is a familiar term, as modern humans are referred to as *Homo sapiens*, 'intelligent man'. When we speak of early *Homo*, we mean the first signs of our group in particular – species that adopted a way of life we would recognise as human-like. It is contentious who belongs to this group but generally, species found after the appearance of stone tools in the fossil record (2.5 million years ago) and which are close to our line of ancestors are called *Homo*. The three species that appeared relatively early are *Homo habilis*, *H. rudolfensis*, and *H. erectus*. Some scientists classify the first two as *Australopithecus*, depending on how they choose to distinguish *Australopithecus* from *Homo*, but in any case, when you see the names '*habilis*' and '*rudolfensis*', they refer to the species described here.

Towards humanity

An elderly individual from the Homo erectus site of Dmanisi, Georgia – the oldest hominin site outside of Africa.

Homo habilis

When Louis Leakey and his colleagues announced the discovery of *Homo habilis* in 1964, based on fossils from Olduvai Gorge, Tanzania, they described it as a biped and a toolmaker, with a larger brain than *Australopithecus*. In short, it could be considered the first human-like hominin. One of the people involved in the initial study of *H. habilis* – Phillip Tobias – describes this species as crossing a significant threshold in human evolution, possibly possessing the rudiments of language, another human feature. The name '*Homo habilis*' means 'able and skilful man' and is tied directly to the use of stone tools.

Since its discovery, *H. habilis* has often been controversial, first because scientists felt it was not distinct enough to be a new species, and later because it may combine specimens from two different species. Also, some scientists question whether *H. habilis* really behaved in a human-like way and whether the name *Homo* is justified. Nonetheless, *H. habilis* is distinct from *Australopithecus* and *Paranthropus* species in its brain size and other features, and it is the first hominin species to range from eastern to southern Africa, at sites dated from 2.3 to roughly 1.6 million years ago. A key question, of course, is "Who made those tools?"

Oldowan stone tools

The first stone tools made by hominins are found 2.5 million years ago. They were not sophisticated in today's terms, but to make them still required human adaptations of the hand.

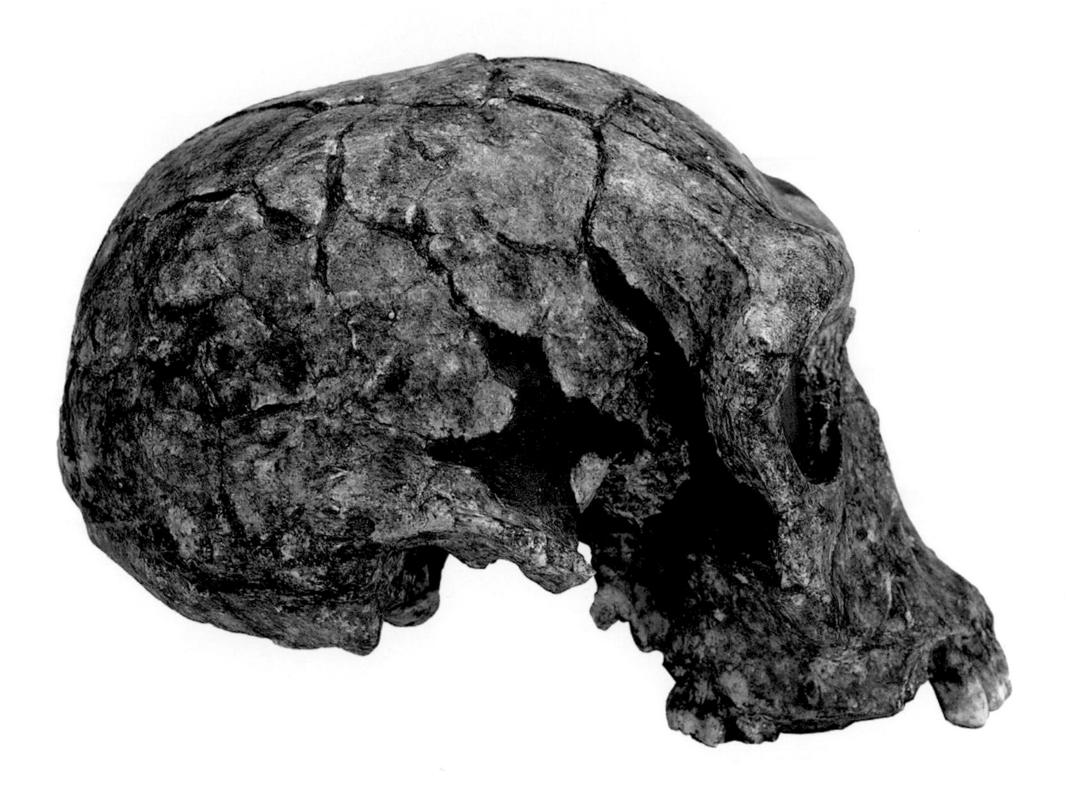

Larger brain

A skull of *Homo habilis* from the Koobi Fora site in Kenya. Relative to its size, *Homo habilis* had a slightly larger brain than species of *Australopithecus*.

***Homo habilis*, toolmaker?**

Tools are a critical part of human life, and we cannot imagine life without them. As modern humans, we are surrounded by the products of our tools. If, 2 million years ago, you observed a group of *H. habilis* individuals taking apart an antelope carcass, their use of tools might remind you of your local butcher's shop, even though the hominins looked completely different from humans today. Ever since Louis and Mary Leakey found the first *H. habilis* fossils with stone tools at Olduvai Gorge, the discussion of this species has been wrapped up with discussion of early toolmaking.

It is not just that the fossils are found with stone tools that suggests that early *Homo* made the tools. Otherwise, we would routinely describe other hominins in the same terms. For example, *Paranthropus boisei* is found at the same sites as *H. habilis*. Perhaps we'll never know exactly who produced the first tools, but the fact that *H. habilis* is on the main human line suggests that they, at least, were doing it. Biologically speaking, *H. habilis* shows slightly larger brain size compared to its overall size. They were small creatures, about the same size as Lucy and other *Australopithecus* individuals, but their brains were a bit bigger.

Toolmaker

Homo habilis has long been discussed as the first maker of stone tools, but now a variety of species are found at stone tool sites.

The hands of *H. habilis* were not entirely human-like, but they show some adaptations to making and using tools in a human-like way. We can use what is called a 'precision grip', meaning simply that we can press our thumb and fingertips together with a great deal of strength. This is important in making and using most kinds of stone tools. Although apes and monkeys can grip things between their thumb and fingers, they generate much less force when they do so. The human hand has evolved the ability to grip things precisely and forcefully, so that stones can be rapidly and efficiently used to make sharp flakes and more sophisticated shapes.

When a bonobo (a relative of the chimpanzee) named Kanzi was trained to make stone tools, he was able to learn the process but then found it was easier to simply smash stones on the ground to make sharp flakes. Yet, if making stone tools is part of your daily life, and crucial to obtaining food, a more efficient method is necessary. In this respect, *H. habilis* lived up to its name – its hand bones show a firm precision grip. Making tools required not just the intelligence but also the physical ability to do it effectively.

What's for dinner?

Tools serve all kinds of purposes for humans today, but we generally think of stone tools as having one key purpose originally – getting meat and marrow from bones. Since hominins lack the kinds of teeth that carnivores use to slice meat off bone (think of dogs, cats, lions, and hyenas), we use tools to do it. You might be surprised how efficiently someone with a few sharp flakes of stone can strip a skeleton of its meat. Early stone tools were also used to break apart limb bones and get at the bone marrow within.

Collections

Archaeologist Fidelis Masao with stone tools from Olduvai Gorge.

If the evolution of *H. habilis* is linked to stone tools, then it is linked to a change in diet. In addition to the staple foods (such as fruit) that probably all hominins ate when possible, *H. habilis* was able to add larger quantities of meat to its diet. Because harder vegetable foods were less important, the chewing muscles of *H. habilis* were smaller than in *Australopithecus* species and much smaller than in the *Paranthropus* species that lived in the same regions. Early *Homo* tooth surfaces in general also had a slightly more edged or crested shape than in other hominins, and this tooth shape would have helped to chew meat.

When we picture humans eating meat, we may also picture hunting, since that is the way most humans obtain animal meat (other than using domesticated animals, a much more recent phenomenon). Whether or not *H. habilis* and other early *Homo* species were hunters has spurred much debate in human evolutionary research. At some level, hunting is not a unique human feature at all. Chimpanzees hunt occasionally, and they prize meat when they can get it. Perhaps all hominins were killing

A new diet

Even the most basic stone tools transformed the hominin diet to include more meat.

and eating small animals from time to time. In any case, *H. habilis* was not hunting with the kinds of weapons that human hunters use today, and their key advantage may have been how fast they could get the meat after the animal was dead, rather than how well they could hunt.

Scavenging is often thought of as a possible option for *H. habilis*. This would give quite a different image of the species – one that would follow on the kills of larger predators and scavenge meat. Exactly how this would work is unclear, because humans and chimpanzees will usually not eat meat that has been lying around for too long, since it would potentially cause illness. While we don't know the full answer to this question, many scientists would say that *H. habilis* may have hunted small animals and occasionally scavenged when it was able to steal a fresh kill (see also *H. erectus* behaviour on pp. 69–77).

Regardless of how *H. habilis* got its prey, there is another implication of eating meat and bone marrow that affected the evolution of this species. Carnivores tend to have wider ranges than animals that eat mostly fruits, leaves, nuts, and

other plant foods. Hunting or scavenging animals successfully usually requires searching over a larger territory. This may be why *H. habilis* is found over a broader area of Africa than *Australopithecus* and *Paranthropus* species.

The little hominin that could

This description may give the impression of a very human-like creature, and indeed that is the heart of the discussion about *H. habilis*. How like us were they? In appearance, *H. habilis* would not stand apart in any dramatic way from their immediate predecessors, the *Australopithecus* species. They were bipeds and the same size as *Australopithecus* – about half the height of a modern human (possibly with larger males, depending on how other fossils are classified – see *Homo rudolfensis*, p. 67). Enormous differences exist between humans today and *H. habilis* then, but this species also reveals the origin of several human behaviours, especially the way that they used stone tools. What *H. habilis* shows, once again, is that human evolution was not a single major event. We did not become 'human' all at once.

Homo rudolfensis

Homo rudolfensis is another of the provocative but less well-known species in human evolution. Defining it is a complicated task. For many years after *Homo habilis* was found in the 1960s, more and more fossils were found and classified as *H. habilis*. Eventually scientists became puzzled at some of the patterns of variation in the growing list of specimens. Some things just didn't seem to fit with our idea of how much individuals can vary within a species. For example, the supposed males and females did not show the same kind of differences that we see in modern apes, humans, and other hominins. Smaller specimens had more prominent browridges – usually a characteristic of larger males – and the anatomy of the face varied more than it does in the species of today. Studying variation in species is an important task, because many of the biological questions we ask about extinct species – where and when they lived, how they behaved, what caused their evolution and extinction – depend on which specimens are classified to those species.

There is still much debate about *H. rudolfensis*, but basically it represents one solution to the problem of too much variation in *H. habilis*. *Homo rudolfensis* are bigger individuals, with larger brains than *H. habilis* and a facial structure that may indicate differences in how muscles were used to chew food. *Homo rudolfensis* is known from a similar time period to *H. habilis* – the most complete fossils are just less than 2 million years old, while more fragmentary ones are between 2 and 2.5 million years old. The species is named after Lake Rudolf, now called Lake Turkana, in Kenya.

Bigger brains – what do they mean?

One of the more variable features of the broad *H. habilis* sample was brain size. There were individuals with *Australopithecus*-sized brains and other individuals that fell in the range of a more human-like species called *Homo erectus*. When put together, the range of variation struck many people as too much for one species, that is, more than you see in chimpanzees, gorillas, or modern humans.

When *H. rudolfensis* is used as a name, it refers partly to the large-brained specimens from Kenya. Does that mean they were smarter than *H. habilis*? Probably not, because the importance of brain size is always relative. Bigger animals have bigger brains just because they are larger on the whole. The skulls of *H. rudolfensis* were also bigger than *H. habilis* in other ways, so relative to overall size, there was nothing special about their brains.

A twist

The problem with describing *H. rudolfensis* as just a big version of *H. habilis* is that they differ in other ways. The face of the most important *H. rudolfensis* specimen is flatter from side to side, and its cheekbone is further forward. Its flatter face looks more like *Australopithecus* than you might expect for a skull that is otherwise more human-like. What this means is unclear, but if *H. rudolfensis* and *H. habilis* were eating different foods, or differed ecologically in other ways, it would help explain why they were able to live in the same region without competing too much for the same resources.

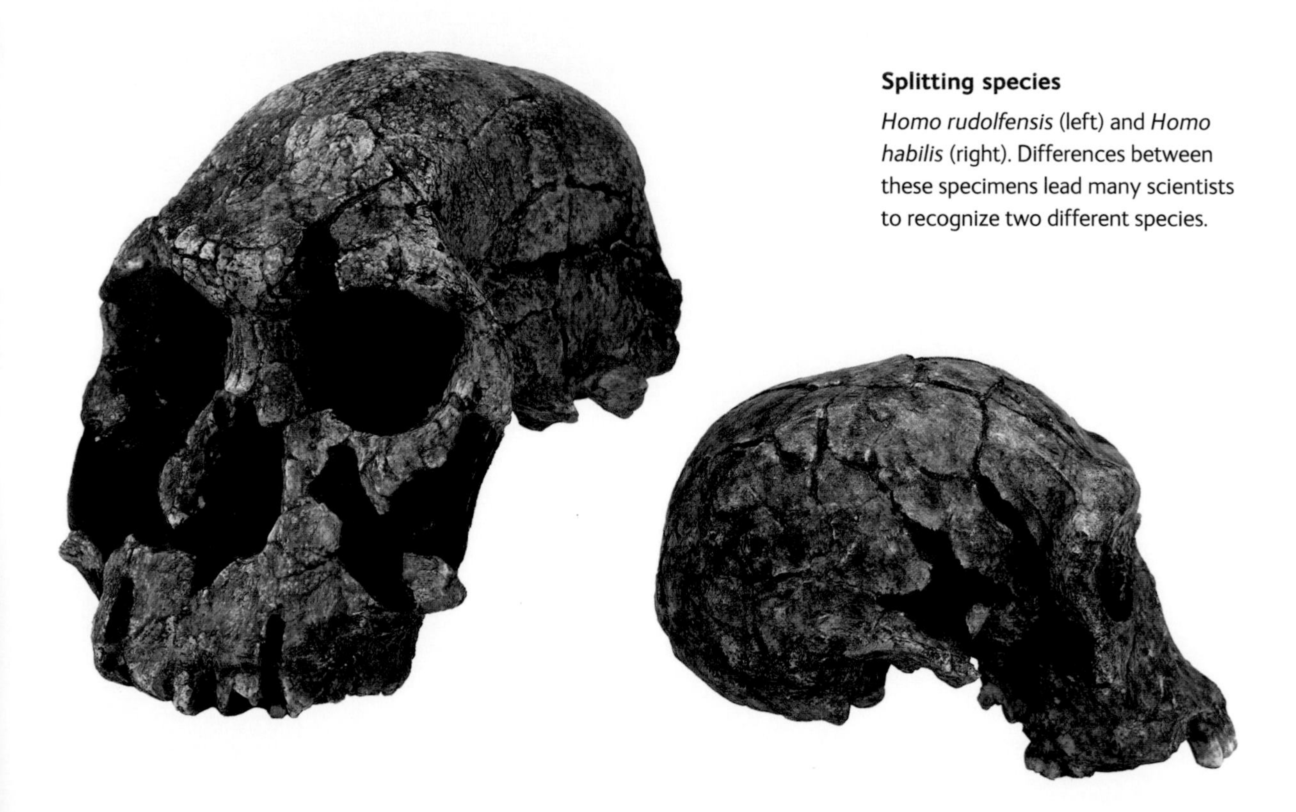

Splitting species

Homo rudolfensis (left) and *Homo habilis* (right). Differences between these specimens lead many scientists to recognize two different species.

An alternative explanation of *H. rudolfensis* is that it simply isn't necessary to recognise different species, or that the current classification is not the right solution to variation in the early *Homo* group. The number of specimens is small, and we still don't understand enough about variation to be sure about this classification. Some scientists continue to interpret the '*H. rudolfensis*' fossils as part of normal variation in a broadly defined species that would be *H. habilis*.

Earliest *Homo*?

Who wins the prize as the oldest specimen of *Homo*? As you can imagine, this is important to know if we want to explain the evolution of *Homo* based on the relationship to another event, such as the first stone tools. The oldest stone tools discovered so far are dated to just over 2.5 million years ago from the Gona site in Ethiopia, and this is an accurate date based on the process of radioactive decay (see pp. 16–17).

Fossil hominins, however, are rarer at 2.5 million years ago (bones do not preserve as well as stone), and they belong to *Australopithecus* or *Paranthropus* species. The best candidates for the oldest member of *Homo* are an upper jaw from Ethiopia (clearly *Homo*, just over 2.3 million years ago), a lower jaw from Malawi (probably *Homo*, possibly as old as 2.4 million years, but the date is less certain), and a piece of a skull from Kenya that mainly contains bones around the ear opening (maybe not *Homo* at all, but the date is over 2.4 million years). Depending on who you ask, these fossils may be attributed to *H. habilis*, *H. rudolfensis*, or *Homo* in a general sense without a species name.

An upper jaw of the genus *Homo*, found at Hadar, Ethiopia, and dated to just over 2.3 million years ago.

It is frustrating not to know the exact answer to this question and to have only a few specimens from this time period. Another problem is that much of what we know about *H. habilis* and *H. rudolfensis* comes from the more complete fossils just under 2 million years old. The earlier fossils may belong to these species, but scientists have not yet confirmed whether they have all of the features we are interested in (e.g. brain size or features of hand bones). If you interpret the record as it stands, stone tools were first made and used by hominins prior to the origin of the genus *Homo*, but we need a better record between 2 and 3 million years ago to resolve the order of events and find out who the first toolmakers truly were.

Homo erectus

What defines success in the evolution of a species? Is it overall population size? Geographic range? Impact on the world? Or is it simply how long the species survives? If the latter is the case, then we have a clear winner in human evolution, and it is not us. *Homo erectus* lived from almost 2 million to roughly 200,000 years ago. Depending on what is called *H. erectus*, the latest sites in Java are even as recent as 50,000 years old. Over nearly a 2-million-year span of time we find evidence of more or less the same species. Some other hominins were successful in this sense, too – *Paranthropus boisei* survived approximately 1 million years, and the exact spans of earlier species may have been similar. But the lifespan of *H. erectus* as a species is truly impressive for the hominins we know well. Although 'success' is a fuzzy term in biology, it is clear that something about the *H. erectus* way of life enabled it to survive a period that included many swings in climate and environment.

Handaxe

A modern replica of a handaxe, along with the debris created when the stone tool is made.

Indeed, *Homo erectus* is often described in terms of a series of achievements – they were the first hominins to spread out of Africa, in many places they used a new type of tool (the handaxe), and they were probably the first hominins to control fire. They were perhaps the first hominins who were not merely skilful in the sense of *H. habilis* making tools, but also able to adapt as humans do. It is always wrong to describe evolution in directed terms, as if *H. erectus* was 'trying' to be human. They weren't. They were very good at being *H. erectus*, and that was enough.

Out of Africa, part one

All hominins before *H. erectus* lived only in Africa. This is true for other early members of the genus *Homo*, *H. habilis* and *H. rudolfensis*. But almost as soon as *H. erectus* appears, it is found in Africa, western Asia, eastern Asia, and the southeast Asian islands of Indonesia. Within a rapid span of time, this species spread further than other hominins had done. Why did this happen?

Animals expand their ranges for lots of reasons. It is worth remembering, of course, that *H. erectus* individuals would not have known where they were going. They were essentially seeking

Spreading out

Homo erectus sites are spread over much of Africa, Asia, and possibly Europe. In some places, indicated by the question marks, sites with stone tools are known from more than a million years ago and were likely made by *Homo erectus*, but there are no hominin fossils that allow us to be sure.

out new places to get food and to have less competition while doing it. And even though 100,000 years is a short time in geological terms, spreading across Asia over this period would not have required much migration in the life of a single person. This was not an epic journey of a single group or population, but instead the sort of gradual expansion of range that one sees in other animals, too.

The reason why hominins had all remained in Africa was largely because their habitats did not stretch across the Sinai region

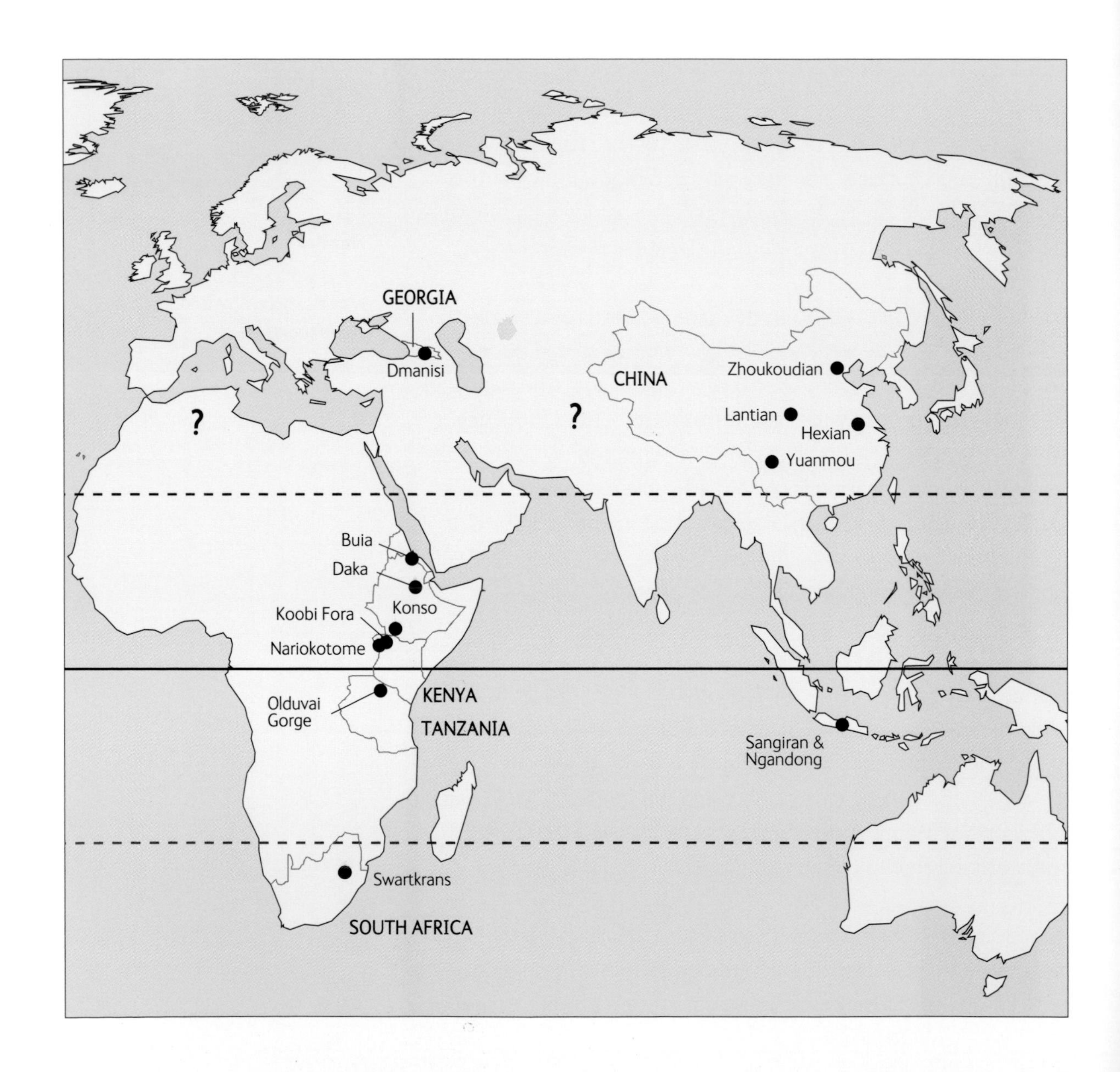

connecting Africa and Asia. Today, this is a hot, dry area, and it is a narrow stretch of land. *Homo erectus* almost certainly did not cross this region under conditions anything like today, but at times in the past the Sinai would have looked different, as local climate fluctuated from more arid to more humid and lush. We can think of groups of *H. erectus* drifting across northern Africa when grasslands, woodlands, and prey animals stretched that far, and some groups would have found themselves crossing into Asia without realising the significance humans would later attach to it.

This event had something to do with how *H. erectus* was living, as well. We know that some hominins had started to make meat more important in their diet even back at 2.5 million years ago. Carnivorous animals such as lions often range more widely than herbivorous animals, because their prey are mobile and have less predictable locations than plants do. Indeed, lions themselves once had a range including parts of Asia and Europe, as well as Africa. *Homo erectus* fits the carnivore pattern, and it was also larger than prior hominin species. An increase to human size, combined with a shift to meat-eating, would explain a wider geographic range, a range that extended out of Africa.

Improvements in technology

It is clear that *H. erectus* did not need new kinds of stone tools to survive at first. The tools found at the earliest *H. erectus* sites are the same kinds seen with earlier species, and *H. erectus* was using this technology when it first spread out of Africa. During the entire lifespan of the *H. erectus* species, however, some things did change. Handaxes are found in Africa after 1.6 million years ago. These were larger tools that required more preparation than the simple choppers that existed before, but they aren't found everywhere that *H. erectus* is.

What we don't know is when exactly *H. erectus* started doing other things differently. The use of fire is perhaps the most important innovation during the era of *H. erectus*, as hominins who controlled fire would have been able to cook food. Cooking is important not just for meat but also to make tough vegetables easier to eat. Vegetable foods – whether raw or cooked – are just as important as meat for human populations living in tropical climates.

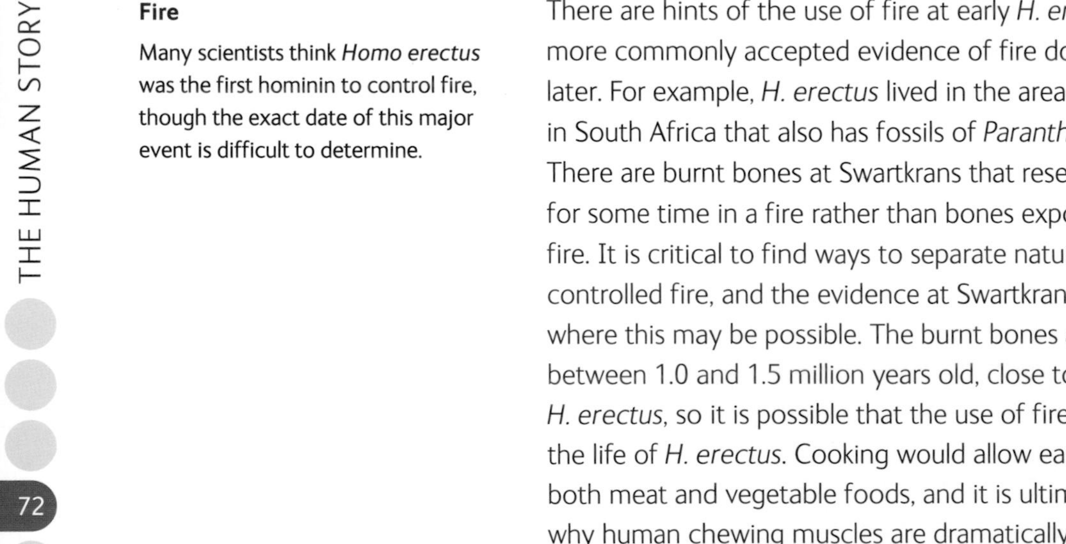

Fire

Many scientists think *Homo erectus* was the first hominin to control fire, though the exact date of this major event is difficult to determine.

There are hints of the use of fire at early *H. erectus* sites, but the more commonly accepted evidence of fire does not come until later. For example, *H. erectus* lived in the area of Swartkrans, a site in South Africa that also has fossils of *Paranthropus robustus*. There are burnt bones at Swartkrans that resemble bones cooked for some time in a fire rather than bones exposed to a quick brush fire. It is critical to find ways to separate natural fire from human-controlled fire, and the evidence at Swartkrans is the earliest point where this may be possible. The burnt bones at Swartkrans are between 1.0 and 1.5 million years old, close to the origin of *H. erectus*, so it is possible that the use of fire was always a part of the life of *H. erectus*. Cooking would allow easier consumption of both meat and vegetable foods, and it is ultimately the reason why human chewing muscles are dramatically reduced compared to those of our ancestors.

Life in Asia

Although we describe *H. erectus* spreading out from Africa to Asia, the discovery of fossils happened in reverse. *Homo erectus* has always been known as an Asian species, from the first discoveries by Eugene DuBois at the Trinil site in Java, Indonesia, in the late 19th century. And it is in Asia that *H. erectus* survived for so long. In Africa, *H. erectus* evolved into species described in the next section of this book, such as *Homo heidelbergensis*. In Asia, *H. erectus* remained much the same as it was initially. Scientists often refer to Asian *H. erectus* as 'classic *Homo erectus*', since these discoveries defined the species.

Sangiran, Java, Indonesia
Some of the first discoveries of *Homo erectus* in Asia were made here.

Asian *H. erectus* fossils are very robust – the bones of the skull are thick, and features such as browridges and other ridges on the skull are pronounced. The brain size was less than that of modern humans but greater than *H. habilis*, *Australopithecus*, or *Paranthropus*. These 'robust' features are not the same as the robust features used to describe *Paranthropus* species (see pp. 46–57).

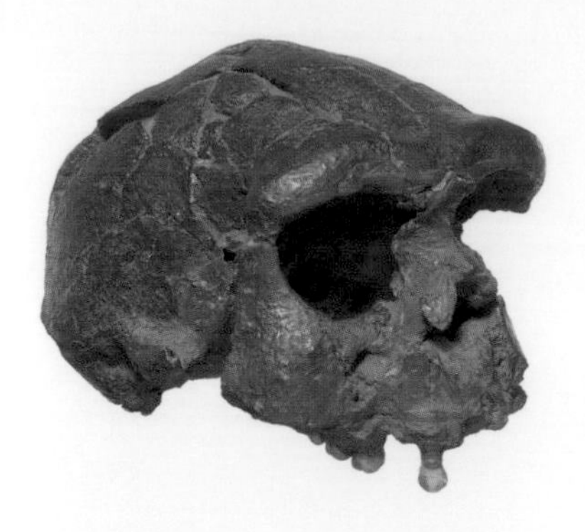

Indonesia

A *Homo erectus* cranium from the Sangiran site in Java, showing the exaggerated browridge and thick skull bones.

For reasons to do with their physiology, growth patterns, and overall level of activity, *H. erectus* individuals had thick bones, but the shape of the skull, for example, was broadly human-like rather than adapted to a diet requiring massive chewing muscles.

Some of the youngest evidence of *H. erectus* – 200,000 to 500,000 years in age – comes from an extensive set of specimens from Zhoukoudian in China, 50 kilometres southwest of Beijing. This is a famous site, and the first finds in the 1920s and 1930s there gave rise to the nickname 'Peking Man'. Sadly, many of the best fossils were lost during the Second World War, and only casts remain. Work still continues at Zhoukoudian, providing new discoveries.

China

One of the localities at the Zhoukoudian *Homo erectus* site in China. *Homo erectus* survived here long after *Homo heidelbergensis* evolved in Africa.

Investigating the scene

If you want to know how species such as *H. erectus* and *H. habilis* behaved, you have to look for a certain kind of forensic evidence. Using tools to take apart a skeleton and scrape off the meat leaves a characteristic pattern of damage and marks. The marks are 'cut marks', literally the marks left by stone flakes when bones are cut apart or scraped. Such marks are common on animal bones found at places like Olduvai Gorge, and there is something remarkable about looking at the work of a butcher from nearly 2 million years ago. The patterns on the bones indicate that *H. erectus* (and probably *H. habilis*, too) was able to get at the best joints of meat.

Behaviour

Homo erectus showed many similar behaviours to modern humans, though the extent of human-like language in this species is unknown.

Marks are found, for example, close to the hip and shoulder joints of the animals they ate (a variety of antelope species and occasionally some larger mammals such as hippos). If *H. erectus* had been a scavenger, these thicker portions of meat would have been long gone by the time it had its turn. Instead, the pattern of marks suggests that *H. erectus* was either hunting the animals itself or immediately stealing kills from other predators, as modern human hunters sometimes do. Either way, this meat butchering and eating was a significant part of its behaviour.

Complete skeleton

The 'Turkana Boy' skeleton from Nariokotome, Kenya, is the most complete individual of *Homo erectus*.

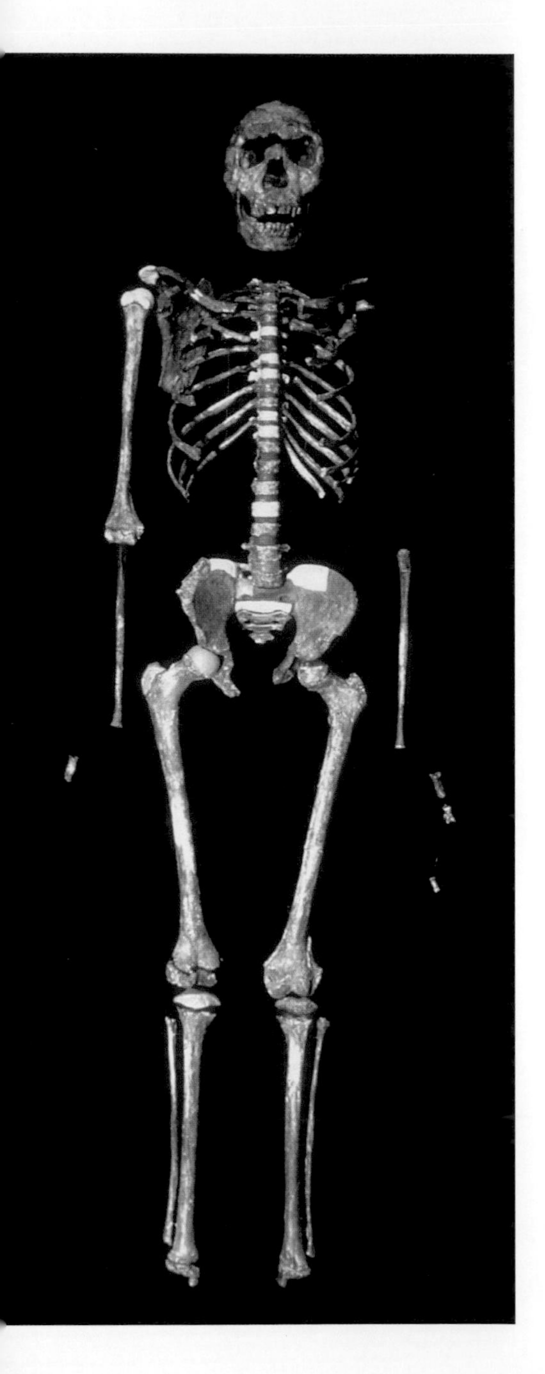

A growing boy

The discovery that has most influenced our understanding of *H. erectus* in recent times is the Nariokotome boy. Nariokotome is a site in the Lake Turkana region of Kenya that features frequently throughout the human story. In 1984, a team led by anthropologists Richard Leakey and Alan Walker discovered one of the most complete skeletons that has been described for any hominin before more recent species such as the Neanderthals. Hip anatomy shows that he was a male. The Nariokotome boy was only at the stage of growth of a 12-year-old human, yet he was 160 cm tall (5 foot 3 inches). This height is similar to tall humans of that age and taller than earlier hominins. His skeleton is also human-like, and he is thought to have walked much as we do.

Unlike early hominins such as Lucy (see pp. 22–30), *H. erectus* probably did not use trees for safety or as a main source of food. Climbing was a rare activity, as in modern humans – useful sometimes for hunting or gathering foods, but not the critical way of getting around that it was for apes and early hominins. We think *H. erectus* lived in a range of habitats, including more open habitats like modern savannas and grasslands. They may have been safer in these habitats than earlier hominins, because they were larger or if they were able to control fire. The Nariokotome boy gives us an image of *H. erectus* as a tall, human-like species, walking across the landscape as humans do today.

Another very human feature of Nariokotome is that he had skeletal problems. In particular, he suffered from curvature of the spine (scoliosis) and had a hunched posture. His own individual story is just as interesting as the implications for *H. erectus* as a species.

A more complicated pattern

Discoveries in recent years have revealed more variation in *H. erectus*, showing an unexpected and more interesting pattern. The site of Dmanisi, in the Republic of Georgia in western Asia, astounded scientists through the 1990s and early 21st century. They discovered a series of well-preserved skulls that looked like *H. erectus* and yet did not show all of the expected features. In particular, some of them have small brain sizes similar to the range found in *H. habilis*. The skulls from Dmanisi are all around 1.8 million years old, which is at the beginning of the time when *H. erectus* lived. This makes us wonder how different the *H. erectus* populations were from each other, and it also shows that the first hominins to leave Africa were more similar to *H. habilis* than we thought. Further discoveries may show that *H. habilis* itself spread out of Africa.

First stop

The Dmanisi early hominin site in the republic of Georgia was discovered beneath the ruins of a medieval city.

Homo ergaster

Homo ergaster is the name sometimes given to African specimens of *H. erectus*, and it can be used for the newer discoveries from Dmanisi in Georgia as well. The name 'ergaster' comes from the word 'workman' in its Greek form, and it was named this way for reasons similar to the naming of *H. habilis*. The African fossils are found with stone tools, and in some cases these are more advanced tools than those used by *H. habilis*.

Early African fossils of *H. erectus* do not have all of the 'classic' *H. erectus* features associated with finds first made in Asia. The simplest description is that they are less robust, and this overall effect is seen in a number of correlated features of the skull. Scientists who separate *H. ergaster* from *H. erectus* generally think that the African species is on the human line, and that the Asian specimens of *H. erectus* eventually went extinct rather than evolving into other forms. The impressive Nariokotome skeleton also belongs to *H. ergaster* if this name is used.

Some scientists refer to African *Homo erectus* fossils as a different species, *Homo ergaster*, because they do not have all of the features seen in Asian populations of *Homo erectus*.

Regardless of how populations are named, after 1 million years ago *H. erectus* was no longer found in Africa. Although *H. erectus* survived until relatively recently in Asia, the African populations became something different, the subject of the next section.

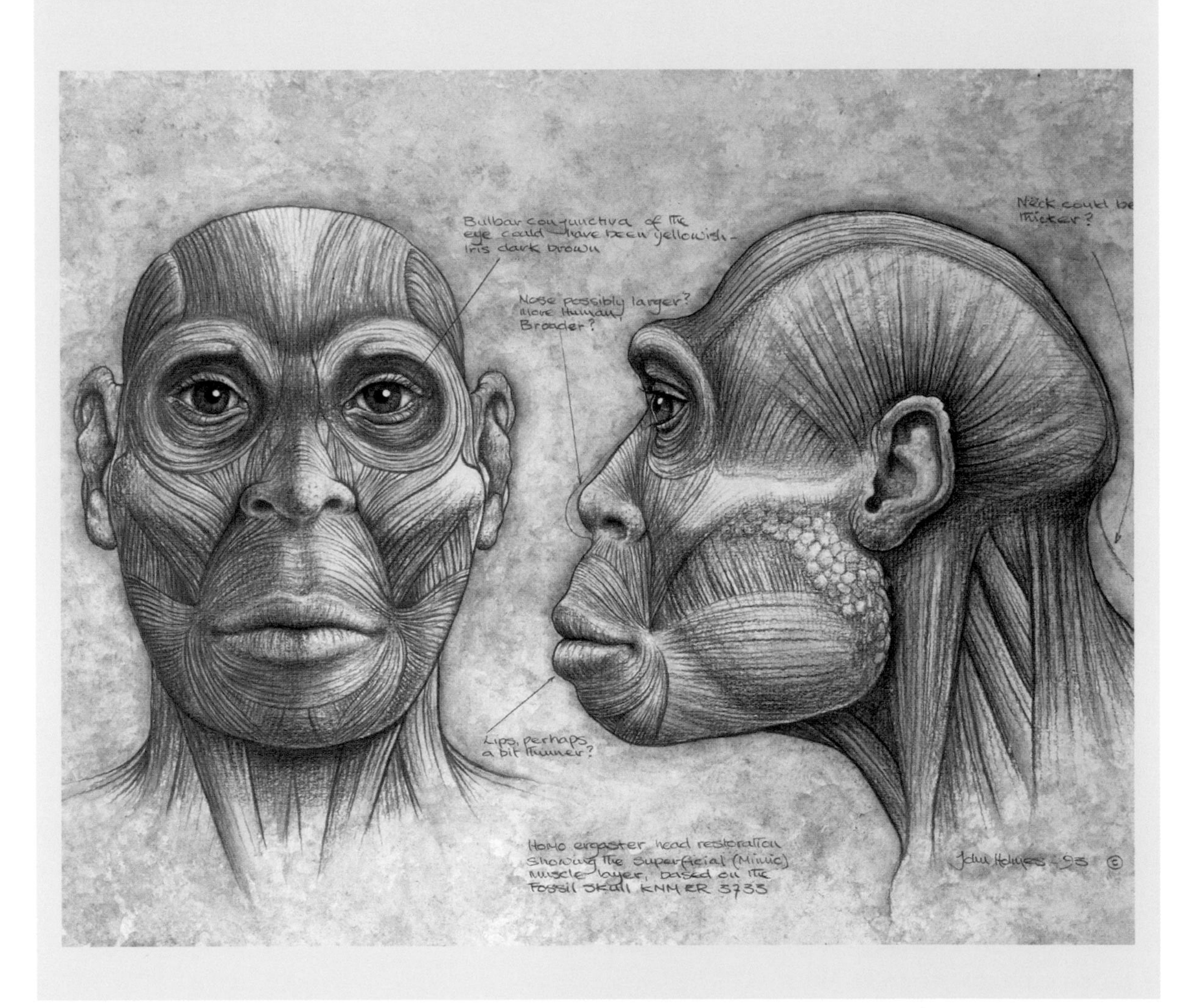

THE HUMAN STORY

Differences also exist between the African and eastern Asian specimens of *H. erectus*, and for this reason African specimens are sometimes called *Homo ergaster* (see box on p. 78). When we study a hominin that lived on several continents, we are faced with complex evolutionary questions that do not apply to earlier hominins. *Homo erectus* reveals the challenge of studying the past – the more we know, the more questions we ask, and the more limited the evidence seems. Despite being relatively well known, and surviving such a long time, *H. erectus* is still full of surprises.

Discovery in context

One of the skulls from Dmanisi before it was removed in 2001.

LATER *HOMO*

After one million years ago, we see hominin species beyond Africa, living across Asia and in parts of Europe. By this point in human evolution, we think of these hominins as distinctly human. To some scientists, all of the species described here as 'later *Homo*' belong to our own species, *Homo sapiens*. If that is the case, they are all populations within an evolving species, one that changes over time and shows some regional variations but is essentially one genetically cohesive group. A few scientists would go as far as including *Homo erectus* in *H. sapiens* as well. This book adopts the more common view – that the recent fossil record of human evolution includes a variety of species. One implication of recognising species diversity is that not all of the species are in the direct line of human evolution. The modern human lineage is just one branch of the hominin tree.

Old England

Excavation at the 500,000 year old site of Boxgrove, one of the most important sources of evidence for early hominin life in Europe.

Homo heidelbergensis

Some time before 700,000 years ago, a group of hominins found themselves on a warm subtropical coast, with long summer nights and short winter days. They were surrounded by other animals adapted to warm climates, such as elephants and hippos. Those hominins didn't know that they were living in what would eventually be called England, known for quite different weather. At that time, the world was in a warm phase, and many tropical animals – including our ancestors – had spread north as far as northern Europe. The site of Pakefield on the eastern English coast is the oldest evidence of humans living in the British Isles, but the evidence consists only of their tools. The species that made the tools is a bit of a mystery, but it was possibly *Homo heidelbergensis*, a species known from Europe and Africa somewhat later in time.

Human arrival in Britain

A flint tool found at Pakefield, the earliest hominin site in Britain.

The name '*Homo heidelbergensis*' has been around since it was first coined for a jaw found near Heidelberg, Germany, in 1907. But the concept of *H. heidelbergensis* is essentially a new and changeable idea. It is most easily thought of as a descendant of *Homo erectus* in Africa, which spread out to Europe. In this way *H. heidelbergensis* was the ancestor for Neanderthals and modern humans. However, as the box on p. 87 about *Homo antecessor* and *Homo rhodesiensis* shows, there are lots of names for the hominins that lived in *H. heidelbergensis* times, and the picture could be much more complicated.

If we assume that only one species existed in Africa and Europe between 300,000 and 700,000 years ago, *H. heidelbergensis* is the name of that species. It is recognisable from its human-like but very robust and massive skull. Some individuals had brain sizes falling in the modern human range of 1200–1500 cubic centimetres (about three times as big as *Australopithecus* brains). Skulls of *H. heidelbergensis* from places like Greece, Spain, Ethiopia, and Zambia are remarkably similar to each other, which is why the species is thought of as both European and African. As time passes, however, European populations begin to look different, giving rise to the well-known Neanderthals.

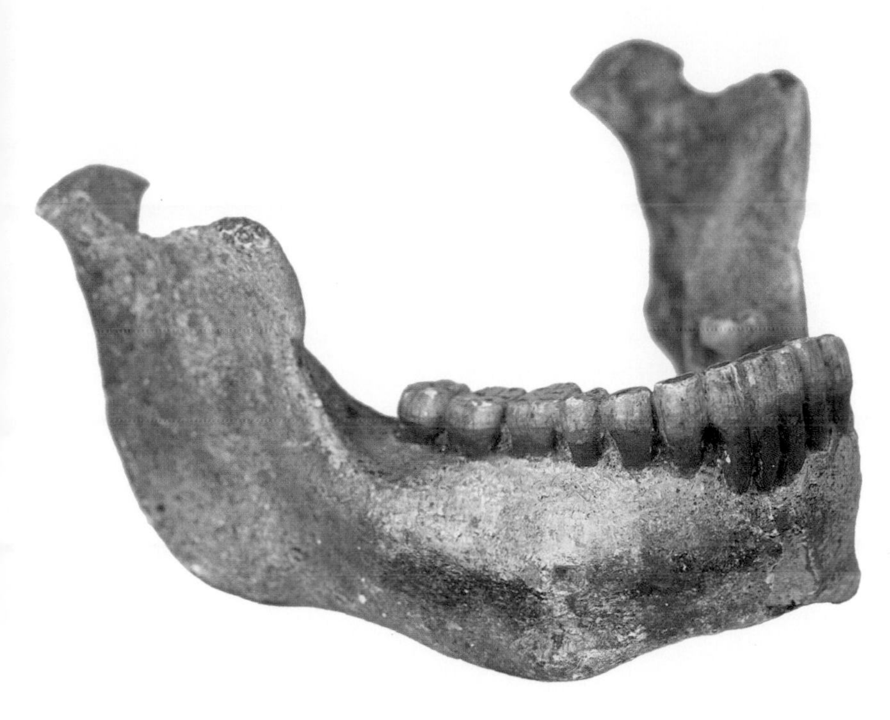

The oldest spears

Scientists often debate the evolution of hunting behaviour. This is relevant to technology and archaeology, to cooperation among individuals, to what the hominins ate and to how far they ranged. *Homo heidelbergensis* gives us dramatic evidence of hunting, mainly in the form of wooden spears. These spears are about 400,000 years old and were found in Germany. They are the oldest recognisable weapons (tools used for hunting rather than butchering) in the human fossil record, and horses found in the same site were probably the targets.

First specimen

The name *Homo heidelbergensis* is based on this lower jaw from the Mauer site, near Heidelberg, Germany.

Before Neanderthals

A *Homo heidelbergensis* skull from Petralona, Greece, showing the basic human shape but more robustly built.

As extraordinary as they are, these spears also remind us of what we do not see in the fossil record. Wooden tools are only preserved in the rarest of circumstances – the archaeological record is almost entirely focused on stone and bone. The wooden spears in Germany are relatively simple, sharpened poles, without stone tips. Weapons like these might have existed in much older times, without ever being preserved. In any case, they tell us that *H. heidelbergensis* was hunting large animals and in this sense was human-like.

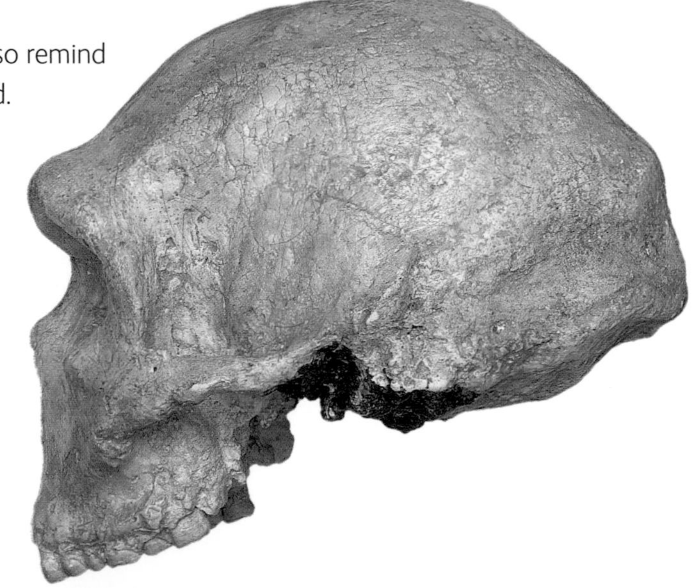

What is an 'archaic human'?

Homo heidelbergensis is known mostly from skulls, except for the unique collection of skeletal remains from the Spanish site of Atapuerca (Sima de los Huesos), which is discussed more in the section on Neanderthals (pp. 88–97). Bones of the skeleton give the impression of a human-sized, heavily built, bipedal hominin. Muscle markings are more pronounced than in most modern humans, and limb bones have thick layers of hard bone around their central marrow cavities. Reconstructions often portray these people with 'bodybuilder' physiques. These aspects of size are important. Humans are the second biggest modern primates, after gorillas. A lot of resources are needed to maintain such size, and *H. heidelbergensis* faced this challenge as much as modern humans.

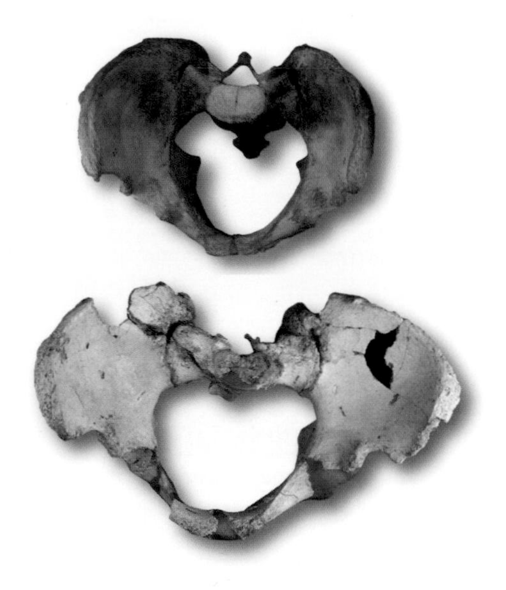

A heavy hominin

One of the most massive individuals in the human fossil record (bottom) is represented by huge hip bones of *Homo heidelbergensis* at the site of Atapuerca, Spain.

Lacking chins, and with a longer, lower shape of the skull, they fit an image of 'archaic', retaining some characteristics of our earlier ancestors. This appearance, rather than brain size, is why the term 'archaic humans' is used to describe *H. heidelbergensis* and some other recent species. Their behaviour was broadly human-like. But if you were to encounter a *H. heidelbergensis* person, you would see them as a different species, especially because their facial appearance was not in the normal, modern human range of variation.

When *H. heidelbergensis* lived, we are still talking about a time when multiple hominin species lived on Earth. It is easy to slip into thinking of human evolution as a linear process, from one species to the next until modern humans evolved. But when *H. heidelbergensis* evolved in Africa and Europe, *H. erectus* continued successfully in eastern Asia (read the section on *H. erectus* to see just how long that species survived). For hundreds of thousands of years there were at least two human-like hominins living at the same time but in different geographic regions. Only one of them gave rise to later humans, and this, we think, was *H. heidelbergensis*.

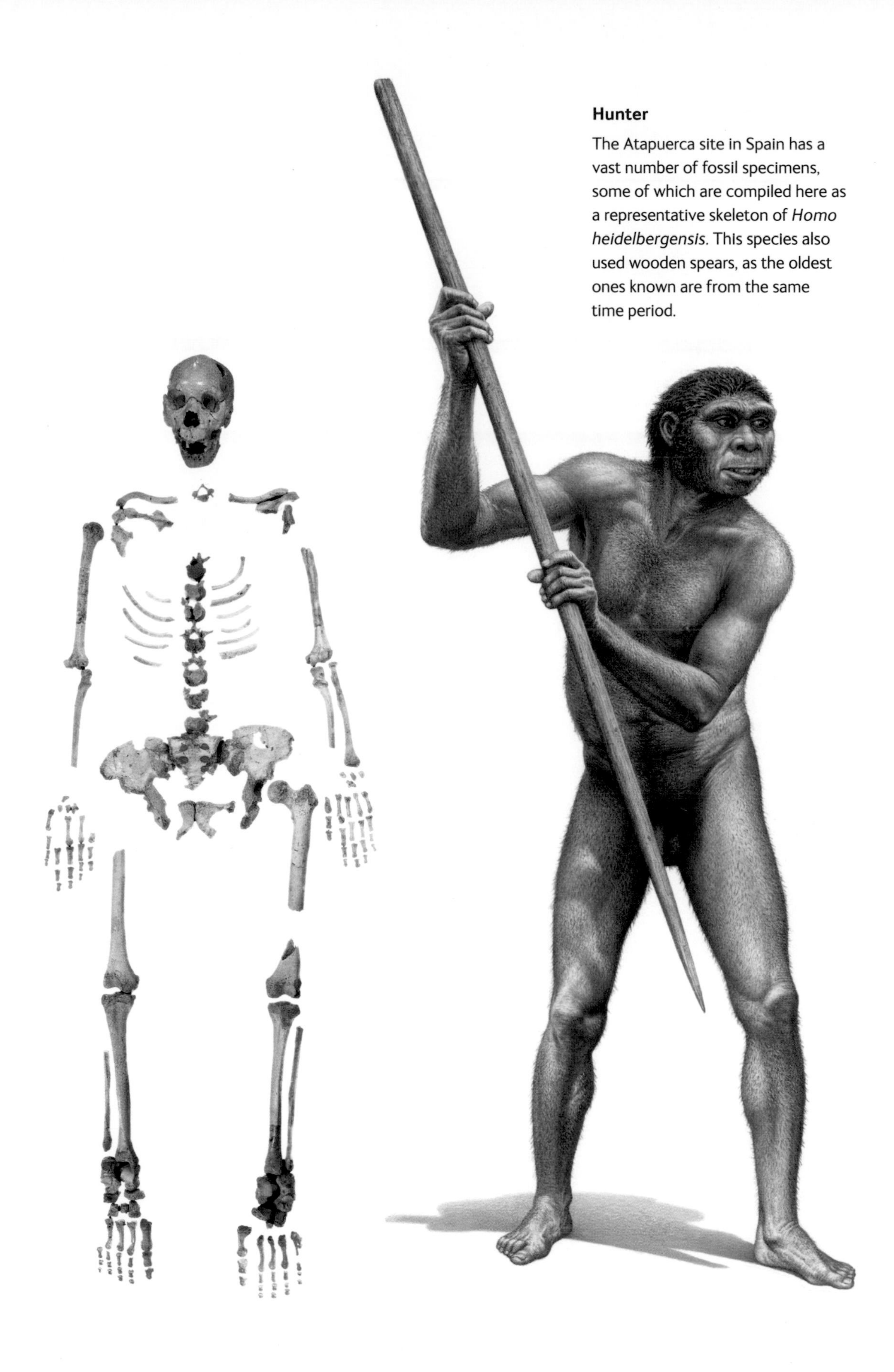

Hunter

The Atapuerca site in Spain has a vast number of fossil specimens, some of which are compiled here as a representative skeleton of *Homo heidelbergensis*. This species also used wooden spears, as the oldest ones known are from the same time period.

The 'archaic' face

Homo heidelbergensis individuals lacked chins and had much larger browridges than modern humans.

Climate change

The warm and hospitable climate of northern Europe during *H. heidelbergensis* times was not permanent. This is obvious from the present climate, but even hundreds of thousands of years ago there were periods when Europe became much colder than it is now. During the Pleistocene – a geological term for the epoch stretching from 1.8 million years ago up to 10,000 years ago – Earth had begun its swings between ice ages and warm stages, or, more formally, between glacial and interglacial phases. As time went on, the swings became more and more extreme. The changes in climate may have forced *H. heidelbergensis* to migrate across the continent to track its prey and to avoid the harshest areas.

The cold periods may also have affected the anatomy of *H. heidelbergensis* in Europe, as natural selection favoured individuals better able to survive the low temperatures. This phenomenon is well illustrated by Neanderthals, the species that followed *H. heidelbergensis* in Europe. Because the latest European specimens we call *H. heidelbergensis* are different in many ways from the earlier specimens, and more like Neanderthals, how to divide the two groups is unclear. In a sense, the last *H. heidelbergensis* in Europe were also the first Neanderthals.

Homo antecessor and *Homo rhodesiensis*

The period between early *Homo erectus* and modern humans may seem to be a simple pattern, with *Homo heidelbergensis* filling the role of a 'transitional' species. But a glance through descriptions of human evolution will reveal a variety of names that suggest multiple lineages at this time. All of these names come down to the question of what species are and how much difference between populations is enough to mark them as different species. For the period between 300,000 and 1 million years ago, two other hominin names that you might see are *Homo rhodesiensis* and *Homo antecessor*.

Homo rhodesiensis was originally coined as a name for a specimen from Zambia. Scientists who think that archaic African populations were a different species from European populations use *H. rhodesiensis* for the African species and *H. heidelbergensis* for the European species. Other scientists just call them all *H. heidelbergensis*. Anatomically and behaviourally, *H. rhodesiensis* was similar to *H. heidelbergensis*.

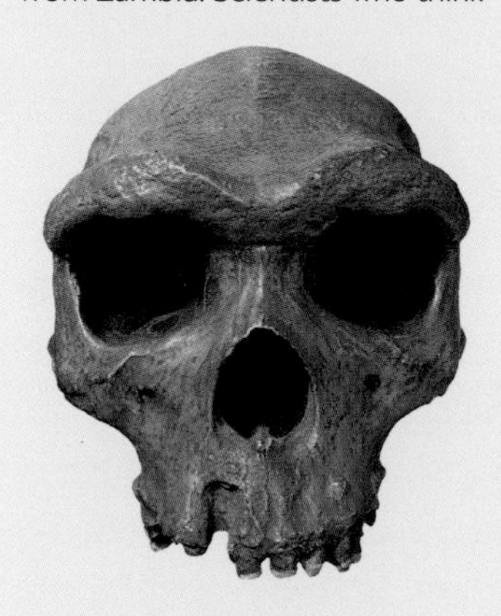

A skull from Zambia, sometimes assigned to the species *Homo rhodesiensis*.

Excavations in Atapuerca, Spain.

Homo antecessor is a more complicated story, based on the oldest fossil hominins found in Europe. The mountainous northern Spanish site of Atapuerca is famous not only for its remarkable collection of *Homo heidelbergensis* fossils from more than 400,000 years ago (see pp. 82–86), but also for the discovery of European fossils, dated to 800,000 years. The most informative specimens are mainly parts of the skull of a teenager. They were called *H. antecessor* because of differences in facial anatomy between them and other hominins – especially the presence of a 'sunken' area below the cheekbone called the canine fossa, also a human feature. Not many fossils are known from this time, so it is unclear how many species there were. Some scientists have called other early Europeans *H. antecessor* as well, but for simplicity many refer to all of them as *H. heidelbergensis*.

Until there is solid agreement on the details of human evolution in the last million years, the names may change somewhat.

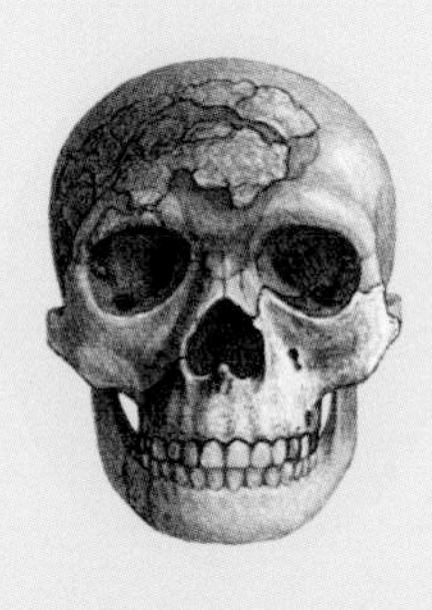

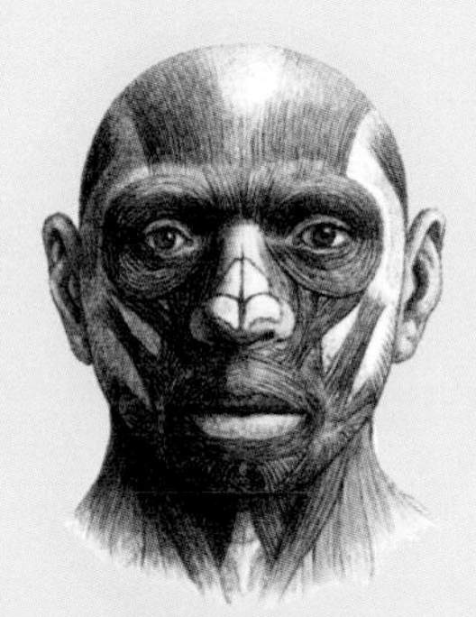

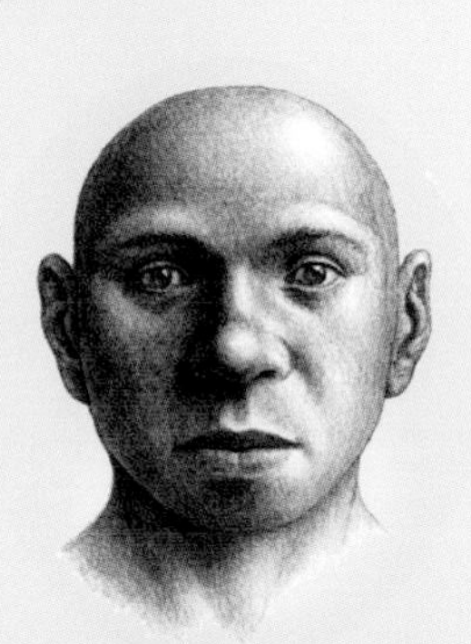

***Homo antecessor* was named based on a different facial appearance from *Homo heidelbergensis*.**

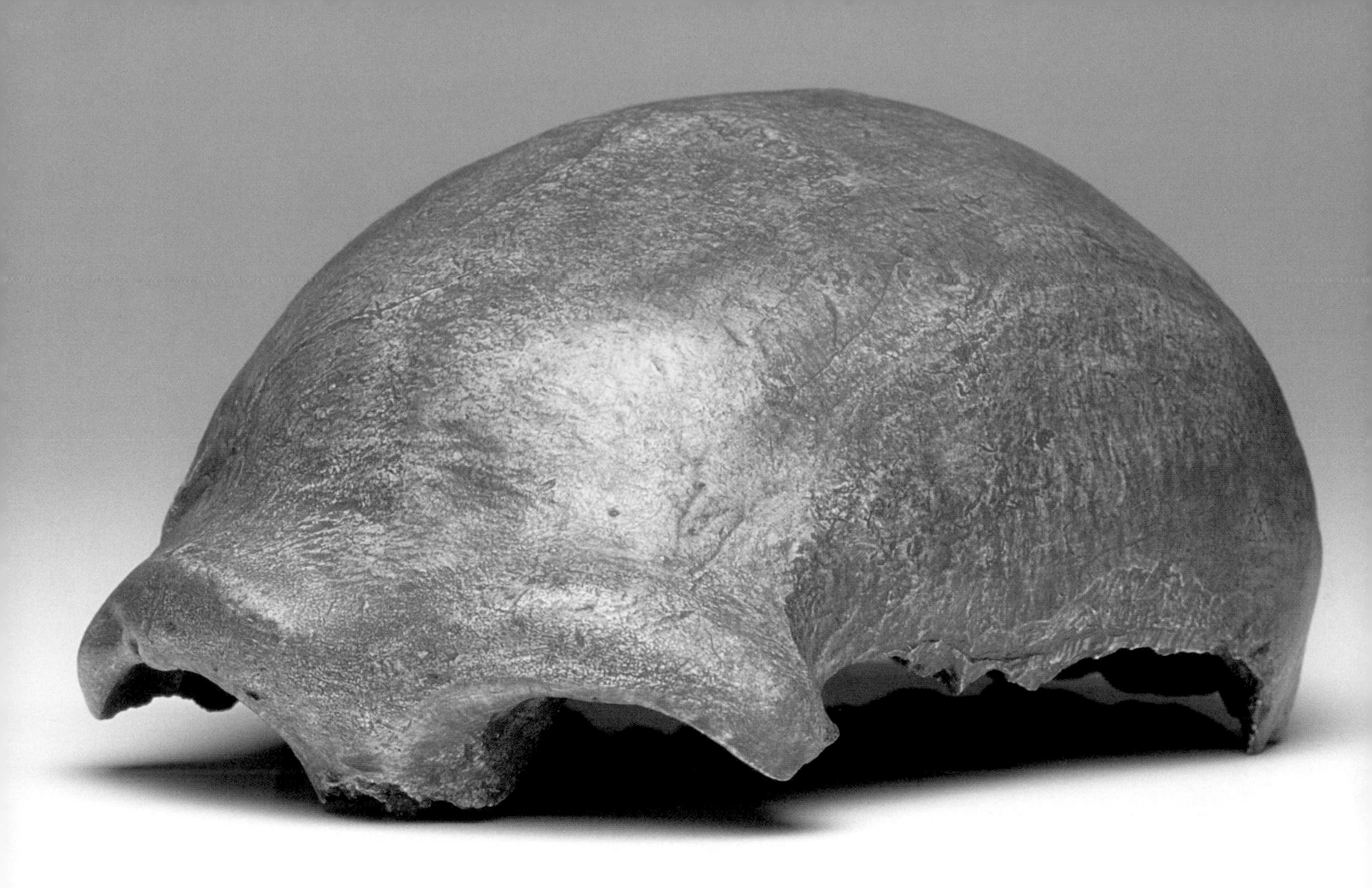

Homo neanderthalensis

Revolutionary find

The first specimen of a Neanderthal, discovered with parts of a skeleton in the Neander Valley, Germany, in 1856.

When you hear 'Neanderthal', you probably think of it as a broad term for human ancestors. In reality, *Homo neanderthalensis* is the best-known hominin species from the past and specifically a European and west Asian group. Ironically, it is also one of the greatest mysteries. We know an extraordinary amount about the Neanderthals' behaviour, where they lived, and when, but scientists still debate whether they are an extinct hominin species or a population that is part of the modern human lineage.

The image of a Neanderthal is familiar – a projecting face, no chin, a large browridge, and a human-sized brain in a skull that was longer from front to back than a human skull. But what do these features mean? How different were Neanderthals from humans today?

Brawn and brains?

Some have described Neanderthals as using brawn more than brains to solve their problems. As with *H. heidelbergensis*, muscle markings and bone thickness indicate a powerful physique. This physique was normal for most of human evolutionary history. No doubt Neanderthals were strong people, but this does not mean they had limited intelligence. In fact, their brains were as large as a modern person's brain. In a broad sense, Neanderthal behaviour was not that different from other hominins living at the same time in other parts of the world. They hunted, and they may have hunted more than other populations. Europe was cold during most of Neanderthal existence, and in cold climates humans rely more on animals for food. Neanderthals had also mastered the control of fire.

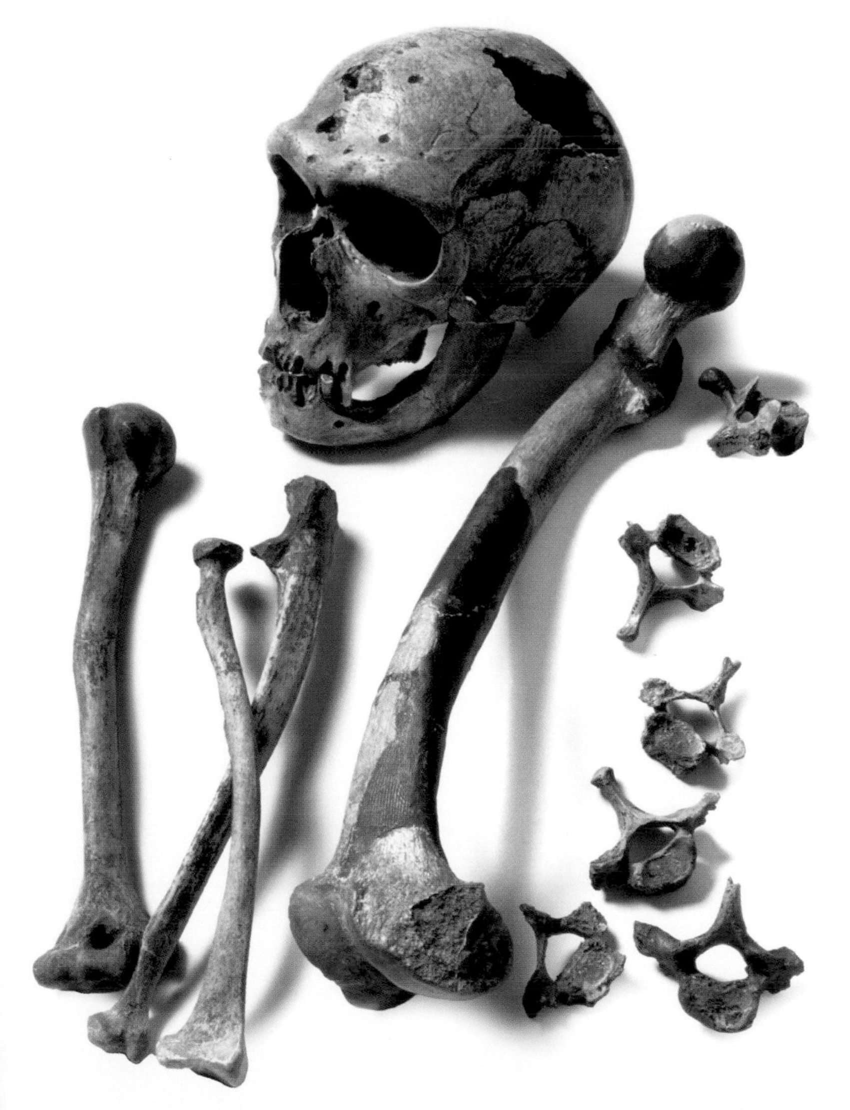

Signs of age

The 'Old Man of La Chapelle', a skeleton of an elderly Neanderthal from about 60,000 years ago in France.

Neanderthal tools

Neanderthals used new, more efficient techniques to produce stone tools referred to as 'Middle Palaeolithic' or 'Mousterian' tools. These were critical in carving meat from bison, reindeer, and other large mammals.

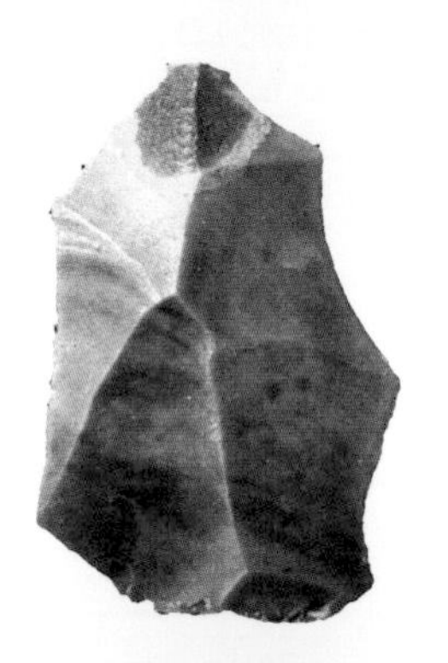

What is particularly interesting about Neanderthals is the amount of damage visible in their skeletons. Despite their strength and their robust bones, Neanderthals show high rates of broken bones and dislocated joints, along with dental diseases and arthritis. The bone damage and arthritis show the amount of stress they put on their skeletons. Neanderthals were not 'superhuman' in any sense. It is likely that many injuries came from hunting large animals at close range – Neanderthals preyed on bison and horses as well as smaller animals like red deer and gazelles. Neanderthals used spears but did not have technology such as bows and arrows or spear-throwers to help them kill their prey from a long distance.

In many respects, Neanderthals were like other hominins from the last several hundred thousand years, the group of hominins generally called 'archaic humans'. Modern humans today are less muscular and have less robust bones than our ancestors. This is true even for humans living in traditional ways, such as hunting and gathering foods. For Neanderthals, the question is whether they were a distinct species for other reasons.

It's cold up here

When human ancestors first arrived in Europe, it was semi-tropical – a logical place to go if you were a species that evolved in hot Africa (see the section on *Homo heidelbergensis*, pp. 82–86). But, over time, Europe went through 'glacial' periods of extreme cold. The changing climate caused movement of animals back and forth to follow their habitats, but in many cases it also produced evolutionary change. Animals that were more highly adapted to cold conditions were better able to survive. These animals included reindeer, a frequent target of Neanderthal hunting during cold periods. As part of a process of evolution, European hominins transformed from *H. heidelbergensis* to the species we recognise as Neanderthals.

Many mammals show a pattern of adaptation to temperature. Mammals living in higher latitudes (further north or south, in places with colder winters) are better at conserving heat, while mammals closer to the equator are better at losing heat and staying cool when the weather is extremely hot.

How did this affect Neanderthals? We describe Neanderthals as short and stocky, and their relatively short arms and legs are considered to be an adaptation to cold environments. Heat loss in humans is affected by how much surface area the body has, in other words how much skin is exposed to the weather. With short arms and legs, humans become better at keeping heat in. Modern humans that live in the far north, such as some European populations and especially the Inuit of North America and the Sámi populations of northern Eurasia, tend to be different from other humans in this way. Neanderthals show a more extreme form of this adaptation. That said, Neanderthals lived in a broad swathe of territory from western Europe all the way east to what is now Uzbekistan, and south to sites in Israel. It appears that their range may have shifted with climate, and that they weren't capable of handling the most severe temperatures.

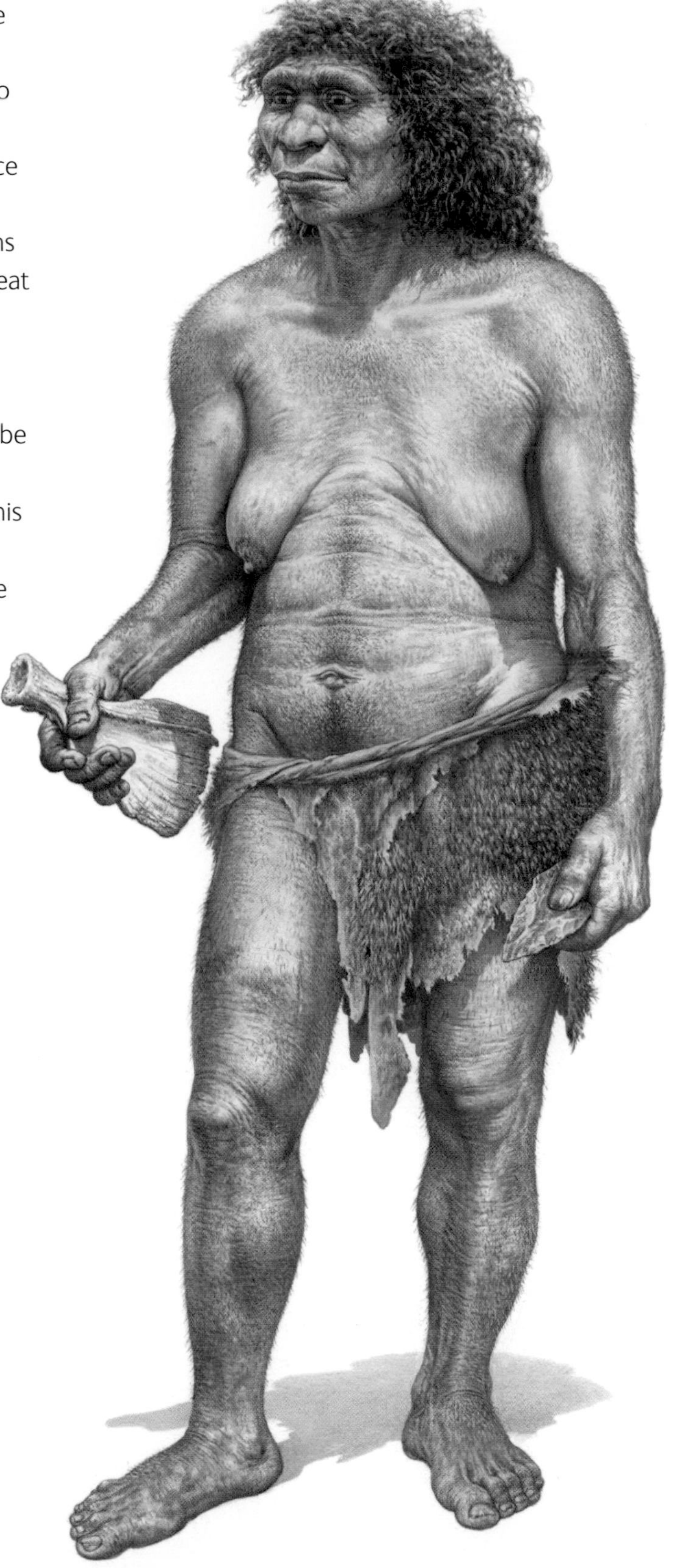

Neanderthal physique

Neanderthals were shorter and stockier than most humans today.

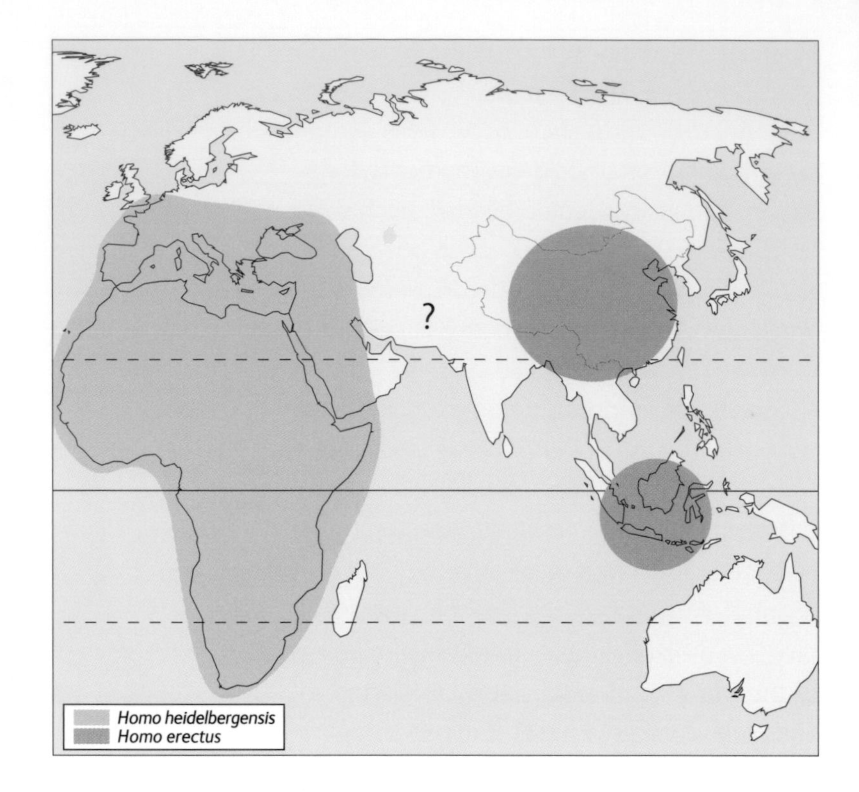

Diverging species

By 500,000 years ago, *Homo heidelbergensis* had evolved in Africa and spread to Europe, but *Homo erectus* still lived in eastern Asia and islands of Southeast Asia.

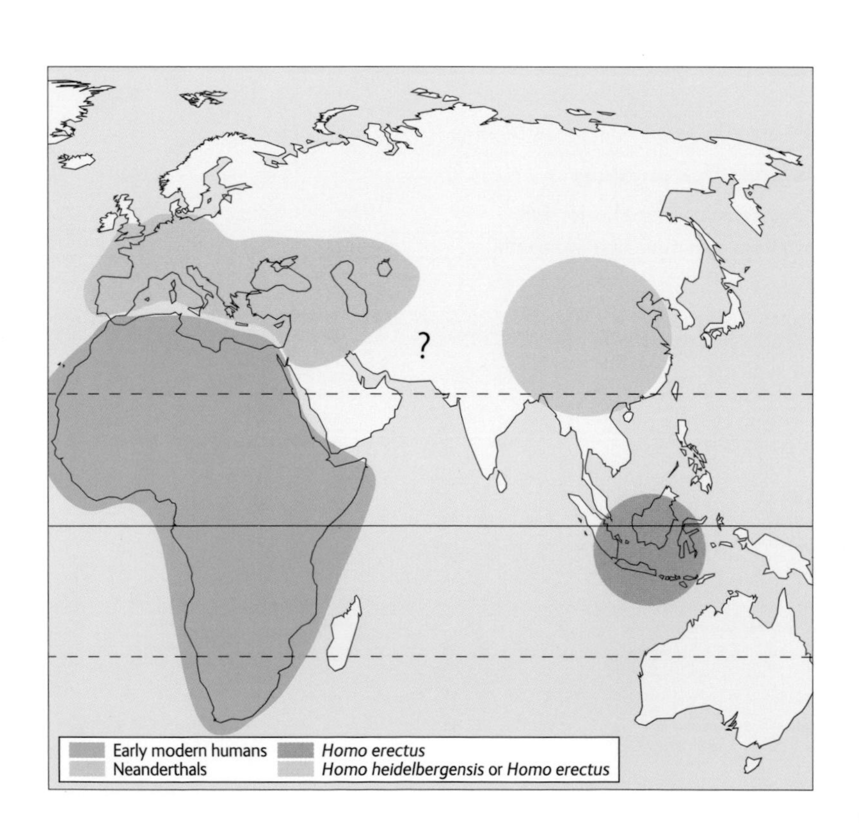

A complex pattern

When modern humans evolved in Africa 150-200,000 years ago, Neanderthals were still living in Europe and western Asia, and *Homo erectus* or other hominin species were living in eastern Asia. After this point, modern humans began to spread out from Africa and replace other hominin species.

Burying the dead

In trying to decide how human-like the Neanderthals were, we need to consider what their social behaviour was like and how their groups were structured. In some ways they were strikingly like us. For example, Neanderthals buried their dead, whether for reasons of honouring them or to maintain hygiene in the places they lived. Some of the most recent Neanderthals also made simple jewellery for themselves, such as necklaces of beads or animal teeth. Were they 'modern' because of this?

The differences between Neanderthals and modern humans have to do with quality and quantity. Early in their history, modern humans used similar tools to Neanderthals, but later in time they made more sophisticated and more diverse tools, showing more delicate techniques (see *Homo sapiens*, pp. 100–107). Modern humans used beads and ornaments earlier than Neanderthals (some scientists conclude that Neanderthals learned these techniques from modern humans), and they eventually expanded their artwork to include cave paintings and a variety of sculpted objects. Either there were simply fewer Neanderthals around – which would make sense in the cold European habitat – or they had less need and interest in the kinds of sophisticated art that modern humans produced. We also come back to the limitations of the fossil record. Ornaments of bone and teeth preserve well in the fossil record, but artefacts of wood or any behaviour that uses plant parts or dyes are more difficult for archaeologists to detect. One of the most famous and debated finds with Neanderthals is the discovery of flower pollen with a burial at the Shanidar site in Iraq. It could indicate the symbolic burial of flowers with a person, or the pollen may have got there by other means.

Early burials

The Kebara Neanderthal skeleton from Israel. Early burials were often simple, without the placing of special objects.

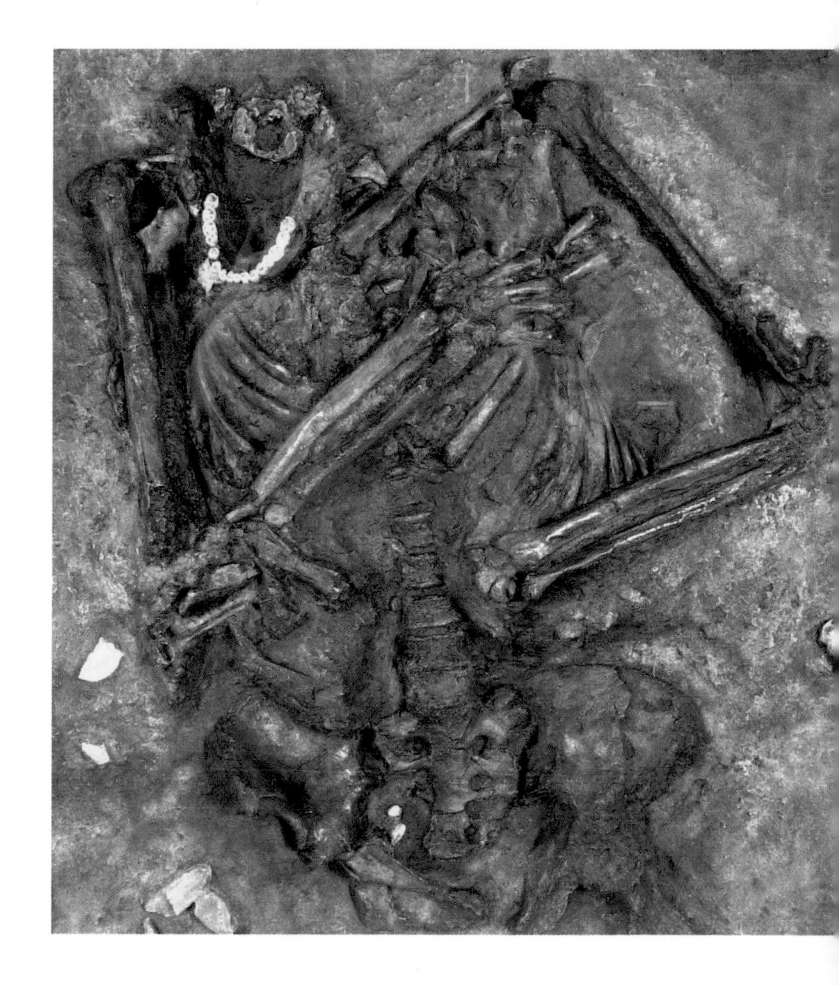

The first Neanderthals

Asking who was the first Neanderthal is like asking who was the first human. Because evolution is a process of change, it is impossible to agree on the point where species become what they are. In a sense, we can 'see' Neanderthals evolving by the appearance of some Neanderthal features at the extraordinary site of Atapuerca, northern Spain, between 400,000 and 600,000 years ago. Scientists call the fossils *H. heidelbergensis*, because they do not show all of the Neanderthal characteristics. For many years, scientists also referred to specimens like these as 'pre-Neanderthals' because of their intermediate anatomy.

By 100,000 years ago, Neanderthals possessed the many distinct features that allow us to recognise them, including large, projecting faces and large noses; a braincase that was human-sized but much lower and longer from front to back; and robust skeletons with curved limb bones. They also have features on the back and underside of the skull that are very rare in modern humans, such as a projecting mound of bone on the back called an occipital bun. Individuals with all of these characteristics are often called 'classic' Neanderthals.

Evolutionary trends

The original Neanderthal skullcap (bottom row, middle) in a line of skulls belonging to - from upper left - *Australopithecus africanus, Homo rudolfensis, H. erectus, H. heidelbergensis,* and, on the bottom right, a modern human. Although it is possible to line up skulls like this through time, the actual pattern of human evolution is more complex.

Atapuerca is one of the few sites that tells us about a group or a population, rather than just a few individuals. It contains a remarkable number of people for a fossil sample, at least 36 of all ages and both sexes. The Atapuerca site was reported only recently, in the early 1990s. Since then, detailed study has made it one of the most important sites in the world in human evolution. Specimens from the site reveal even the smallest aspects of behaviour, such as evidence of hominins using toothpicks to clean their teeth.

Part of us?

Since the middle of the 19th century, when Neanderthal fossils were first found, scientists have debated whether they were the same species as you and me. This is a profound question that

Telling a story

There has been much debate about whether Neanderthals used language. It is likely that they did, even if the way they spoke was different from modern humans.

The Amud cave, Israel

Neanderthals found here are between 50 and 70,000 years old, showing that they lived in the region just before modern humans began a rapid expansion across Asia.

applies to many fossils in the last 1 or 2 million years, but Neanderthals are the most popular focus because they are so recent and the fossils are so numerous. How wide is the human range of variation? What does it mean to have a distinct hominin species living so recently, up to 35,000 years ago? Only about 2000 generations have passed since Neanderthals became extinct. If Neanderthals and modern humans met in Europe, is it possible that humans continue to tell folk legends of Neanderthals? Fossil bones tell part of the story, and they generally show that Neanderthals are well outside the range of bone shape in modern humans. This applies to skulls, limb bones, hand bones, and hips. Genetic evidence is coming into the discussion, too. Analysis of human DNA is rapidly changing the way we think about ourselves, and the complete genetic sequence for modern humans is now known. Because Neanderthals lived recently in time, it is possible to extract some DNA from their fossilised bones (this becomes more difficult the further you go back in time). Recent announcements suggest we may even have a complete genetic sequence for Neanderthals within a few years.

Another burial

An infant Neanderthal skeleton from Amud, Israel. Even at a young age (about 10 months old), we can see Neanderthal features such as the long foramen magnum visible here, and the shape of the lower jaw.

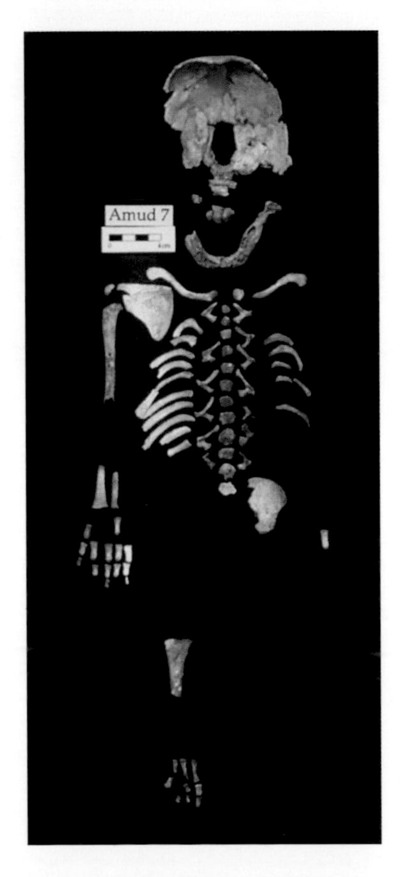

So far, genetic evidence fits the Neanderthal fossil record very well. Humans and Neanderthals seem to have split genetically at about 500,000 years ago, in the era of *Homo heidelbergenis*. Not long after this, Neanderthal traits appear in some European skulls, so it seems that the European (Neanderthal) and African (modern human) populations were evolving separately. There is also no evidence that modern Europeans inherited genes from Neanderthals, as would be the case if Neanderthals were simply a European population within a broader, worldwide species of archaic humans.

Working out exactly whether Neanderthals interbred with modern humans will involve more study of the populations that were close to each other, such as those in western Europe and eastern Europe. But, even if limited interbreeding occurred, it didn't seem to have much effect on subsequent populations. The next hominins in western Europe are distinctly similar to modern *Homo sapiens*, showing that Neanderthals essentially went extinct between 30,000 and 35,000 years ago.

Homo floresiensis

Flores site

The site of the excavation of *Homo floresiensis* in the Liang Bua cave on the island of Flores, Indonesia.

Homo floresiensis is the most recent discovery in the genus *Homo*. If it is indeed a distinct species, it is also the last of our close relatives to have gone extinct. This surprising hominin was isolated on the island of Flores in Indonesia and apparently lived until about 12,000 years ago. It immediately became known as 'the Hobbit' because of its diminutive stature, but this gives a wrong impression of its appearance – the fictional hobbits have large brains, unlike the Flores hominins! It is true, however, that the Flores people were only about 1–1.2 metres (3–4 feet) tall.

Since being announced in 2004, the Flores fossils have aroused much controversy and debate. Their brain sizes were roughly the same as apes, even after accounting for their small body size. The rest of their anatomy shows that they belong in the human family, but their skeletons do not exactly match those of any other species. The Flores population is thought to be an example of 'island dwarfing', a process that happens to many large mammals when they become isolated on islands. Indeed, alongside the Flores fossils are tiny elephants that were a source of food.

Other explanations?

The main hypothesis explaining Flores is that this group started as an isolated population of *Homo erectus* and eventually evolved to a small size with tiny brains. A number of individuals have been found, and they are all small. But only one complete skull is known, and some scientists believe that he or she suffered from microcephaly, a genetic disorder where the brain does not grow normally but remains small throughout life. If so, then brain size in the population as a whole may have been bigger, and the question of whether this group is a distinct species is more complicated. Some argue that both the small stature as well as the brain size can be explained by pathologies that occur in modern humans, and that the Flores fossils are not a distinct species after all.

The Flores population used tools like other populations of the genus *Homo*. Were their brains somehow reorganised to allow them to have tiny brains but still make tools? Some researchers have found differences in brain shape that support this conclusion. Brains are costly organs in terms of the energy they require, and on a small island hominins may have evolved to be more efficient by having smaller brains.

Other aspects of the Flores skeleton are also distinct from modern *Homo sapiens*. Scientists have now studied bones of the arms and legs, and hands and feet, and in each case found patterns of anatomy that are outside the range of variation seen in humans today. In some ways, the Flores skeleton is more similar to *H. erectus* or *Australopithecus*, or indeed unique among hominins. How much of this unusual anatomy is due to evolution in an island environment with constrained resources? Are some of these differences the effect of disease or genetic abnormalities? The Flores discoveries have revealed a number of things we don't know about the recent human fossil record. The number of specimens of *H. erectus*, for example, is very limited except for skulls, and this makes it more difficult to interpret Flores. As more fossils are found, and especially if another skull is discovered, we may have answers to these fascinating questions.

Hobbit-sized

The original skull and jaw of *Homo floresiensis*, with the lower jaw of a different individual of similar size.

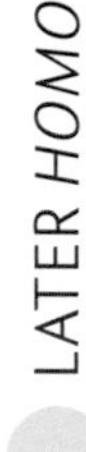

Homo sapiens

What does it mean to be human? There are many answers to this question in human evolution. For an introduction like this one, we can ask what it means to be *Homo sapiens*. The answer to that question defines the beginning of our species in the past, as well as what we are today. It is not an easy task. Because the human fossil record is complete enough to show us a variety of human-like groups in the last million years, there is no single point where the most significant transition took place.

Some would take that to mean that *H. sapiens* is a variable species with a long history. If so, it would include several of the species described in this book, such as Neanderthals. But it is more common to think of *H. sapiens* in specific anatomical terms, as we do with other species. When one does that, it turns out that modern humans have a recent origin.

African origin

A *Homo sapiens* skull discovered in 1967 in the Omo region of southern Ethiopia. The skull is now dated at nearly 200,000 years old.

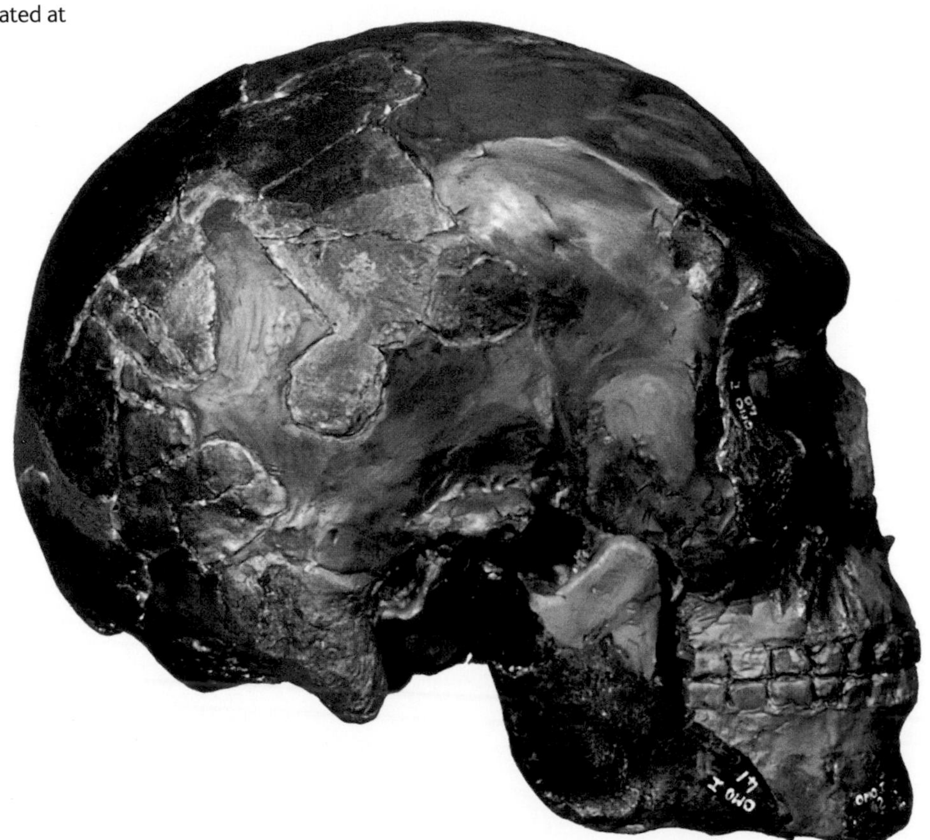

An African origin

Humans evolved in Africa. The very first hominins lived in Africa, and the best evidence for the first modern humans is found there too. In the fossil record, modern humans are referred to specifically as 'anatomically modern *Homo sapiens*'. They are people whose fossil remains are within the range of variation in modern human body shape, or close to it. This phrase does not apply to the Neanderthals, who are anatomically different from us.

The first skulls of *H. sapiens* in this sense are from Ethiopia, from sites between 150,000 and 200,000 years ago. That does not mean that modern humans evolved in Ethiopia first. Where fossils are found depends largely on geology and where it is possible to find them. Modern humans are also found soon afterwards in other parts of Africa, such as South Africa.

The oldest human skulls are recognisable from their chins, the high arching foreheads, the generally rounded shape of the skull, and the characteristic human face. Modern human skulls have a slightly 'sunken' appearance in the face, especially the bones just below the cheekbones. This gives us our prominent cheeks.

Such a portrait is of course just an anatomical description. It is like a forensic anthropologist identifying a person, but we are identifying a species. In their behaviour, the earliest modern humans were not much different from other hominin groups at the time. Their tools were similar, and presumably their modes of hunting, gathering resources, and structuring their social groups were similar as well. When we say 'anatomically modern', we only mean how they are recognised.

Works of art

When then did humans begin to behave the way we do? Many would say that, to understand modern human evolution, behaviour is more important than the anatomy. They suggest we should seek out evidence of humans *thinking* like they do today. This behaviour could express itself as symbolic behaviour, art, burying artefacts with the dead – anything that shows people imagining their place in the world and looking beyond material existence.

Before Picasso

Early modern humans in Europe left behind spectacular cave art, as seen in this mosaic from Lascaux, France.

The most dramatic early works of art by *H. sapiens* are the cave painting sites in France and Spain and 'Venus' figurines found in Austria, France, and the Czech Republic, among other places in Europe. The oldest of these are just over 30,000 years old, and close to the arrival of modern humans in this region. In southern Africa, similar artwork is found from nearly 30,000 years ago as well. The people that made these paintings and sculptures are called 'Upper Palaeolithic', the term used to describe their technology. In a broader sense, the Upper Palaeolithic period began only about 50,000 years ago, long after the earliest modern humans are found in the fossil record. It may be that humans were essentially 'human' long before they showed clear evidence of symbolic behaviour.

On the other hand, such flashy works of art may only be the flourish that came after a long period of this behaviour. More simple ornaments such as beads, and evidence of ochre being used as body paint, go back to similar dates as the first anatomically modern humans. According to this evidence, humans adopted modern behaviour at about the same time as they became anatomically human, and in the same place – Africa.

Where and when

Not long after they appeared in Africa, modern humans spread to the Middle East. Their bones are found there at sites older than 100,000 years ago. This venture out of Africa was brief – climate change seems to have pushed modern humans back into Africa after 100,000 years ago, and Neanderthals lived in the Middle East for some time after modern humans did. Then, around 50,000–60,000 years ago, modern human fossils appear again in the Middle East and as far away as Australia and other parts of Asia. This event is the second major migration out of Africa for hominins. Modern humans spread out to other regions, replacing other hominin populations or occupying lands left empty after those hominins became extinct.

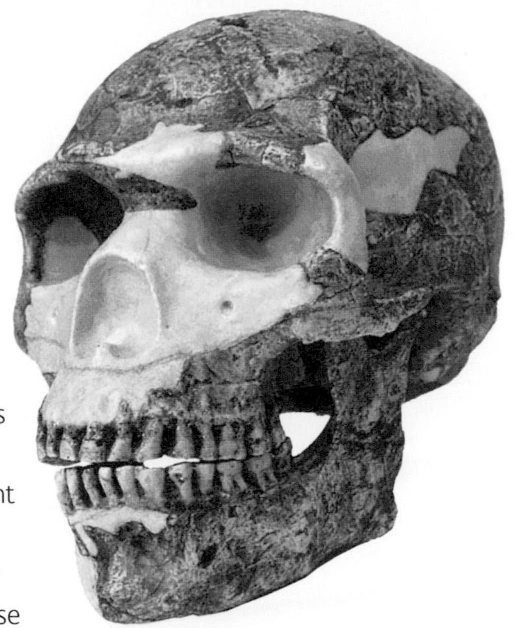

Out of Africa again

A skull of *Homo sapiens* from the Skhul site in Israel, more than 100,000 years old.

Modern human skulls

A selection of skulls of *Homo sapiens*. The two skulls in the middle are among the earliest anatomically modern humans known from Europe (left) and China (right), and the two outer skulls are modern people from the same regions. Recent skulls are often less robust because of changes in lifestyle.

Who were the other hominins? Species such as *Homo neanderthalensis*, *H. erectus* and possibly *H. heidelbergensis* were still around after modern humans evolved in Africa. These other species were living in Europe and Asia. We don't have a clear picture of all these groups, mainly because the fossil record of Asia is poor during the key time period. Neanderthals are known relatively well, and modern humans and Neanderthals lived in different places for a long time. Something seems to have happened 50,000–60,000 years ago to allow modern humans to spread out successfully. We don't know what this was. It may be

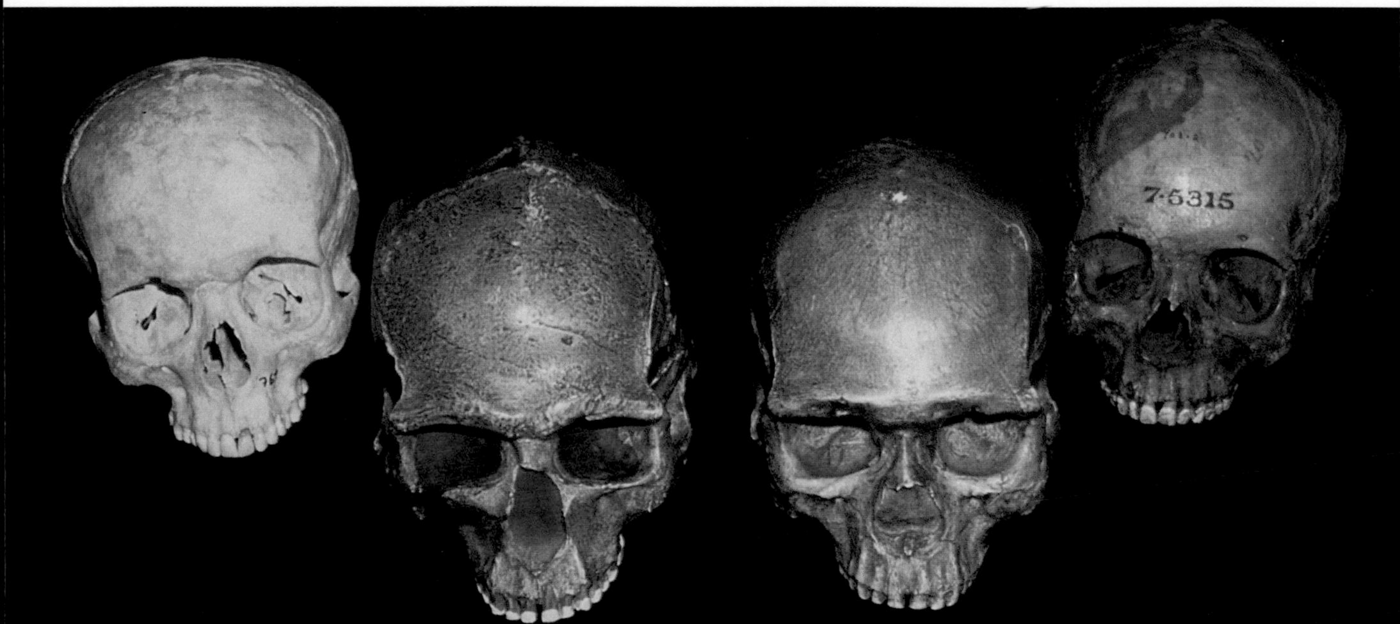

that modern human technology made them more successful in direct competition with other hominins, but it may also be that other hominins did not cope well with climate change and went extinct. Yet another possibility is that language took its modern form at this time, and it allowed the communication of more complex ideas within *H. sapiens* but not in other species.

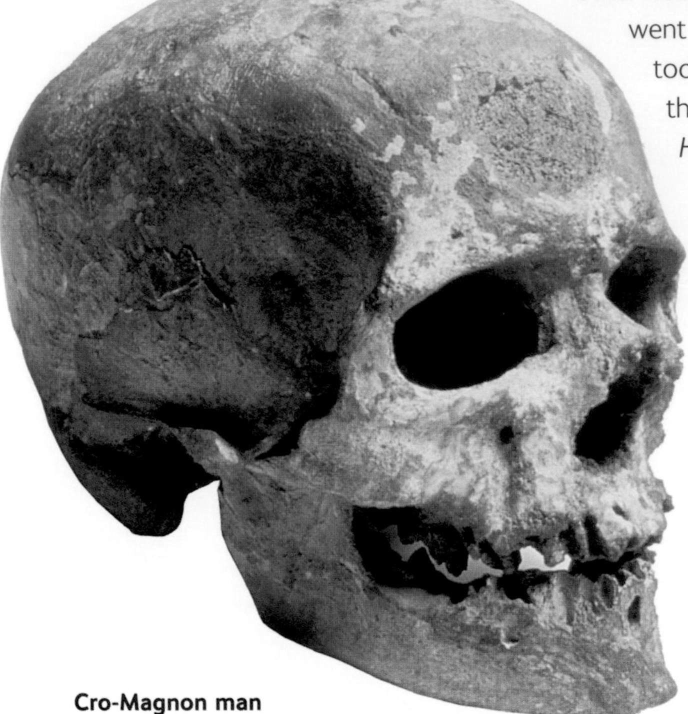

Cro-Magnon man

A skull from the Cro-Magnon site in France. Discovered in the 19th century, this site gave its name to the first modern humans in Europe.

By 30,000 years ago, modern humans had replaced Neanderthals in Europe, and at this point, all other hominin species except possibly *Homo floresiensis* (see pp 98–99) had gone extinct. By 10,000 years ago, modern humans had spread to North and South America. Now, we are essentially in the modern era of the evolutionary record. *Homo sapiens* is unique in having such a broad geographical range, compared to previous hominins or to other mammals. Evidence of a massive expansion of geographical range was seen previously in *H. erectus*, but modern humans truly spread worldwide.

Why?

Humans did not spread throughout the world in order to conquer it. We can still describe these events in evolutionary terms, because population pressure encourages animals of all kinds to spread out when possible. What is different in this case is the wide range of environments occupied by humans and that they replaced other hominins in many places. To a large extent, modern humans were probably more efficient at doing the same things that other hominin species did. In many respects, early modern humans had the same lifestyle as Neanderthals and *Homo heidelbergensis* – hunting animals for meat and gathering a variety of vegetable foods. They were probably nomadic or moved with the seasons. Domestication of animals, the development of agriculture, and more settled communities were a much later occurrence. We don't know whether modern humans competed directly with other hominin populations, but greater efficiency in obtaining food would explain the rapid spread of modern humans

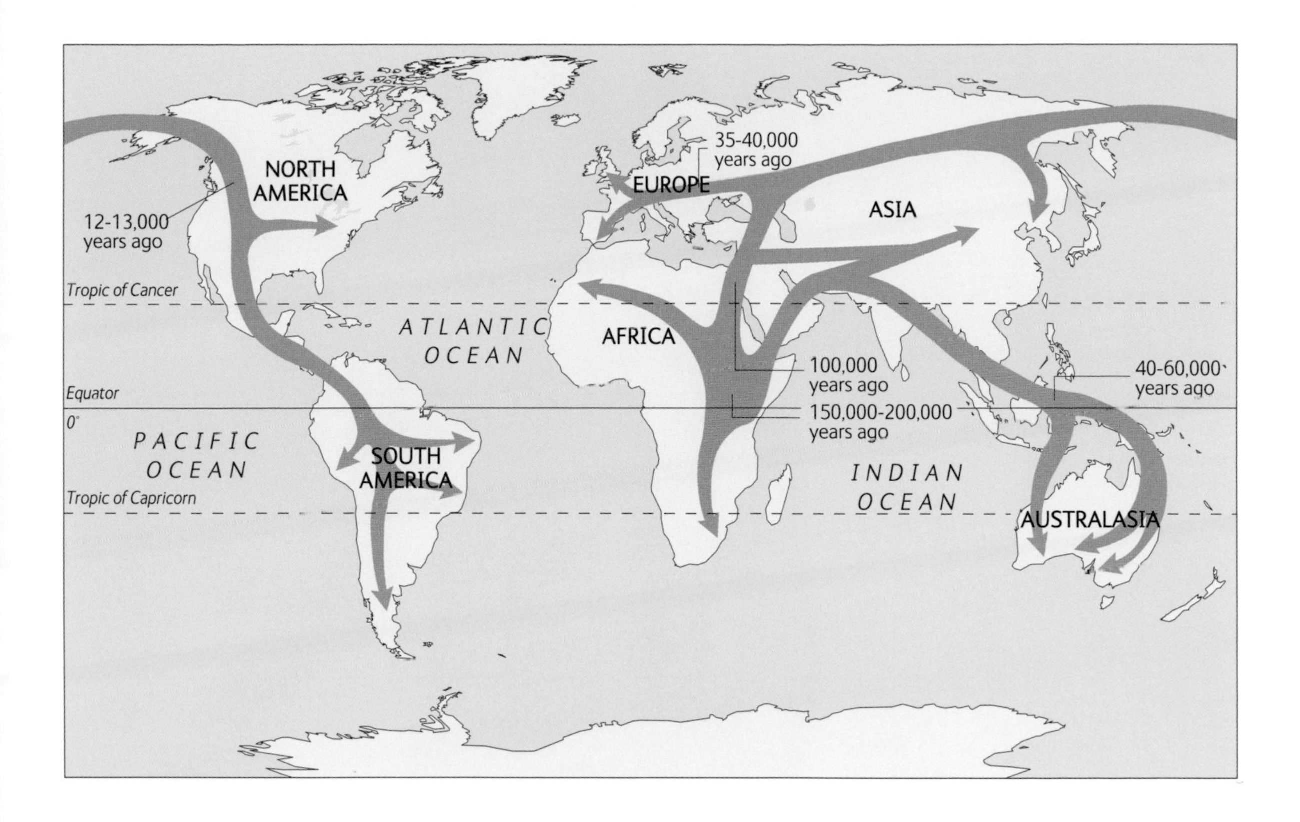

Human dispersal

The first appearance of *Homo sapiens* in the Middle East may have been temporary, as Neanderthals occupied the region after the first modern humans. At 50–60,000 years ago, *Homo sapiens* spread out of Africa again, replacing other hominin populations.

through the world, and our eventual occupation of more extreme habitats than other hominin species.

Along the way, humans developed seemingly endless behaviours, cultures, and languages. We cannot describe in simple terms what humans eat, how they obtain their food, and how they interact with one another as we did for the earlier species in this book, such as *Australopithecus*. It is true that we can describe humans in strictly biological terms, such as bipedal, omnivorous with an emphasis on meat and other cooked food, with complex social structure, and relatively large body size, but this description doesn't capture what we are. Human intelligence interacts with the material world to enable us to live in an extraordinary variety of different ways, through our tools, clothes, and shelter. This great capacity for variation and adaptation is what makes humans human.

Overview – The big picture

After reading this book, you may be bewildered by the number of names used for human ancestors or relatives. These names are essentially the result of pushing the fossil record to the greatest

The last hominin, so far

An anatomically modern human in Europe, from the era of cave paintings and symbolic objects.

limit of detail. But the difficulty with species names is that they are usually coined when new discoveries are made and when the least information is available about a new group. Only over time do we settle on an agreed a way of categorising information about these groups, and it is safe to say that the study of human evolution is still changeable in this sense.

It is important to step away from the names and see the broad pattern, because then we see the remarkable way that the fossil record tells the story of human evolution. It reveals any number of so-called intermediate forms – not ape and not human – which we expect to see in the process of evolution. It shows that the array of features we call 'human' evolved at different times. Human ancestors were bipeds long before brain size increased or stone tools were made. Finally, even the most conservative taxonomist sees diversity of hominin species in the past, especially in the discovery of *Paranthropus* hominins living in the same regions as our direct ancestors. All of these conclusions stem from fossil discoveries, and they represent a consensus that is taught while the finer points of human evolution are debated.

The next step?

Even though we can look back and identify the first signs of human anatomy and behaviour, and describe a process of human evolution, that process was not pre-ordained. It also doesn't allow us to predict what will happen next. With species such as *Homo erectus*, *Paranthropus boisei*, and others, we see how successful some species were for long periods of time. They were not simply transitional groups waiting to evolve into something else. This is true of *Homo sapiens* as well. Anatomically, we are much the same as we were 150,000 years ago. Behaviourally, on the other hand, modern humans show an incredible range of diversity, and technology now evolves much faster than biology. In that shift from biological to technological evolution, the topics of this book give way to recent archaeology and modern history – different subjects but with the same goal of explaining the human past. What comes next is unknown, but we will see the effects of technological change and our own decisions long before humans become something we would call a different species.

Further information

Books

Ape/Man: Adventures in Human Evolution, R. McKie. BBC Books, 2000.

From Lucy to Language, D. Johanson & B. Edgar. Simon and Schuster, 1996.

Human Evolution: A Very Short Introduction, B. Wood. Oxford University Press, 2005.

Human Origins: The Fossil Record. 3rd edition. C. S. Larsen, Waveland Press, 1998.

Principles of Human Evolution, 2nd edition, R. Lewin & R. Foley, Blackwell, 2004.

The Complete World of Human Evolution, C. Stringer & P. Andrews. Thames & Hudson, 2005.

The First Human: The Race to Discover Our Earliest Ancestors, A. Gibbons. Bantam USA, 2006.

The Human Career: Human Biological and Cultural Origins, 2nd edition, R. G. Klein. University of Chicago Press, 1999.

The Last Human: A Guide to Twenty-Two Species of Extinct Human, G. J. Sawyer, V. Deak, E. Sarmiento & R. Milner. Yale University Press, 2007.

Websites

N.B. Website addresses are subject to change. There are many websites that deal with human evolution, but here are a few key sites for your interest.

American Museum of Natural History: http://www.amnh.org

Atapuerca: http://www.atapuerca.com

Australian Museum, Sydney: http://www.amonline.net.au/human_evolution/

BBC: http://www.bbc.co.uk/sn/prehistoric_life/human/human_evolution/

Institute of Human Origins: http://www.becominghuman.org

Leakey Foundation: http://www.leakeyfoundation.org

Natural History Museum: http://www.nhm.ac.uk/

New Scientist: http://www.newscientist.com/channel/being-human/human-evolution

Smithsonian Institution: http://www.mnh.si.edu/anthro/humanorigins/

Index

Italic pagination refers to illustrations.

Picture credits

p.6 © John Sibbick/NHMPL; p.8 © MPFT; p.10 Ben Plumridge © Thames & Hudson Ltd, London, from *The Complete World of Human Evolution* (2005) p185; p.11t © Gerald Eck; p.11b © Institute of Human Origins. Photo: William Kimbel; p.12 © MPFT; p.13t,m Mike Eaton © Natural History Museum; p.13b © Ann Gibbons; p.14 Mike Eaton © Natural History Museum; p.15l,r © Charles Lockwood; p.16 © Sileshi Semaw; p.17 National Museum of Ethiopia, Addis Ababa. © 1994 Tim D. White\Brill Atlanta; p.18 © NHMPL; p.20 © Charles Lockwood; p.22t,m © C. Owen Lovejoy; p 22b © Charles Lockwood; p.23l © John Sibbick/NHMPL; p.23r © NHMPL; p.24 Lisa Wilson © Natural History Museum; p.25l,r © John Reader/Science Photo Library; p.27 © Jay H. Matternes 1984; p.28 © Charles Lockwood; p.30 © Zeresenay Alemseged; p.31 © Koobi Fora Research Project/Meave Leakey; p.32 © National Museums of Kenya; p.33 © W. Schnaubelt N. Kieser-Atelier WILD LIFE ART for the Hessische Landesmuseum Darmstadt (HLMD); p.35 © National Museums of Kenya. Photo: Fred Spoor; p.36 © Don Johanson, Institute of Human Origins; p.37 Lisa Wilson © Natural History Museum; p.38 © Institute of Human Origins. Photo: William Kimbel; p.39 © John Sibbick/NHMPL; p.41 © Mauricio Anton/Science Photo Library; p.42 © Jay H. Matternes 2003; p.44 National Museum of Ethiopia, Addis Ababa. Photo © 1999 David L. Brill\Brill Atlanta; p.46 © Neil Paskin; p.48 © Charles Lockwood; p.49t © Jay H. Matternes 1984; p.49b © Chris Stringer; p.50 © Institute of Human Origins. Photo: William Kimbel; p.51 © NHMPL; p.52 © Fred Spoor; p.53t © Don Johanson, Institute of Human Origins; p.53b © Gerald Eck; p.54t © Don Johanson, Institute of Human Origins; p.54b © Yoel Rak; p.55 © Jay H. Matternes 1999; p.57 © Human Origins Program, Smithsonian Institution; p.58 © Kenneth Garrett; p.60 © NHMPL; p.61 © NHMPL; p.62 © Jay H. Matternes 1999; p.63 © Kenneth Garrett/National Geographic Image Collection; p.64-5 © John Sibbick; p.67l,r © NHMPL; p.68 © Institute of Human Origins. Photo: William Kimbel; p.69 © John Reader/Science Photo Library; p.70 Lisa Wilson © Natural History Museum; p.72 © Jay H. Matternes 1982; p.73 © Chris Turney. *Bones, Rocks and Stars: The Science of When Things Happened.* Published by Macmillan Science, June 2006; p.74t © NHMPL; p.74b © Paul Goldberg; p.75 © Mauricio Anton/Science Photo Library; p.76 © National Museums of Kenya; p.77 © Georgian National Museum. Photo: Guram Tsibakhashvili; p.78 © John Holmes/NHMPL; p.79 © Georgian National Museum. Photo: Guram Tsibakhashvili; p.80 © The Boxgrove Project; p.82 © AHOB/ Natural History Museum; p.83t,b © NHMPL; p.84 © Javier Trueba/MSF/Science Photo Library; p.85l © Javier Trueba/MSF/Science Photo Library; p.85r © John Sibbick/NHMPL; p.86 © Mauricio Anton/Science Photo Library; p.87t © Javier Trueba/MSF/Science Photo Library; p.87m © Human Origins Program, Smithsonian Institution; p.87b © Mauricio Anton/Science Photo Library; p.88 © NHMPL; p.89 © John Reader/Science Photo Library; p.90 © NHMPL; p.91 © John Sibbick/NHMPL; p.92t,b Lisa Wilson © Natural History Museum; p.93 © NHMPL; p.94 © NHMPL; p.95 © Jay H. Matternes 1982; p.96 © Yoel Rak; p.97 © Institute of Human Origins. Photo: William Kimbel; p.98 © Fletcher & Baylis; p.99 © Peter Brown; p.100 © M. Day/NHMPL; p.102 © Pascal Goetgheluck/Science Photo Library; p.103t,b © NHMPL; p.104 © John Reader/Science Photo Library; p.105 Lisa Wilson © Natural History Museum; p.106 © John Sibbick/NHMPL.

Every effort has been made to contact and accurately credit all copyright holders. If we have been unsuccessful, we apologise and welcome correction for future editions and reprints.

(AHOB, Ancient Human Occupation of Britain project; MPFT, Mission Paleoanthropologique Franco-Tchadienne; NHMPL, Natural History Museum Picture Library; MSF, Madrid Scientific Films)